Advances in Experimental Medicine and Biology

Proteomics, Metabolomics, Interactomics and Systems Biology

Advances in Experimental Medicine and Biology provides a platform for scientific contributions in the main disciplines of the biomedicine and the life sciences. This series publishes thematic volumes on contemporary research in the areas of microbiology, immunology, neurosciences, biochemistry, biomedical engineering, genetics, physiology, and cancer research. Covering emerging topics and techniques in basic and clinical science, it brings together clinicians and researchers from various fields.

Advances in Experimental Medicine and Biology has been publishing exceptional works in the field for over 40 years, and is indexed in SCOPUS, Medline (PubMed), Journal Citation Reports/Science Edition, Science Citation Index Expanded (SciSearch, Web of Science), EMBASE, BIOSIS, Reaxys, EMBiology, the Chemical Abstracts Service (CAS), and Pathway Studio.
2021 Impact Factor: 3.650 (no longer indexed in SCIE as of 2022)

Content of this series is reviewed.

This series of volumes focuses on concepts, techniques and recent advances in the field of proteomics, interactomics, metabolomics and systems biology. Recent advances in various 'omics' technologies enable quantitative monitoring of myriad various biological molecules in a high-throughput manner, and allow determination of their variation between different biological states on a genomic scale. Now that the sequencing of various genomes, from prokaryotes to humans, has provided the list and linear sequence of proteins and RNA that build living organisms, defining the complete set of interactions that sustain life constitutes one of the key challenges of the postgenomic era. This series is intended to cover experimental approaches for defining protein-protein, protein-RNA, protein-DNA and protein-lipid interactions; as well as theoretical approaches dealing with data analysis, integration and modeling and ethical issues.

Victor Corasolla Carregari
Editor

Understanding PTMs in Neurodegenerative Diseases

Springer

Editor
Victor Corasolla Carregari
Laboratory of Neuroproteomics, Department of Biochemistry
and Tissue Biology
Institute of Biology, University of Campinas (UNICAMP)
Campinas, Brazil

ISSN 2730-6216 ISSN 2730-6224 (electronic)
Proteomics, Metabolomics, Interactomics and Systems Biology
ISSN 0065-2598 ISSN 2214-8019 (electronic)
Advances in Experimental Medicine and Biology
ISBN 978-3-031-05459-4 ISBN 978-3-031-05460-0 (eBook)
https://doi.org/10.1007/978-3-031-05460-0

This Springer imprint is published by the registered company Springer Nature Switzerland AG
The registered company address is: Gewerbestrasse 11, 6330 Cham, Switzerland

Preface

Neurodegenerative diseases are of great concern for public health around the world; since the population is getting older and life expectancy is constantly increasing, more people are being affected by neurogenerative diseases. It is estimated that almost 50 million people are suffering from a type of neurodegenerative disease, which is also becoming a socio-economic problem and pressuring the public health system as well as increasing manyfold its cost of maintenance.

A better understanding of molecular mechanisms in neurogenerative diseases is a powerful tool to create new treatments and also for early diagnostic. Although there is no detailed pathobiology of neurodegenerative diseases, it is believed that many neural circuits located in different areas of the brain are involved, evidencing that metabolic and signaling pathways regulated by proteins exert an important role in the development of the neurodegenerative diseases.

The majority of these signaling pathways are regulated by acetylation, phosphorylation, ubiquitination, and glycosylation, which makes the study of post-translational modifications (PTMs) essential to obtain a full panorama of what is happening at a molecular level in brain diseases. The regulatory ability of PTMs in various biological pathways and processes is strictly related to addition, removal, or recognition of these PTMs, and changes is these PTMs' sites or in the number of the modified proteins will imply in a dysregulation and malfunction of the pathway leading to disease state. Alterations in PTM states, and their ability to modulate protein-protein interaction have been shown to play an important role in the dysregulation of various biological processes during the brain development hypothesizing that are behind of many central nervous system (CNS) disorders.

The main idea of this book is to increase the awareness of how post translational modifications (PTMs) could be helpful for the understanding of mechanisms in neurodegenerative diseases. In this book, we are aiming to clarify how proteomics studies of PTMs can be carried out in a wide range of neurodegenerative diseases. From sample preparation, PTMs enrichment techniques, and elucidating the role of these

PTMs in neurodegenerative diseases, this book wants to serve as a consultation tool for those who want to dive into the proteomics field and explore the great possibilities that PTM studies can implement in their work.

Campinas, Brazil Victor Corasolla Carregari

Contents

Chapter 1
Protein Extraction and Sample Preparation Methods for Shotgun Proteomics with Central Nervous System Cells and Brain Tissue

Victor Corasolla Carregari

Abstract Shotgun proteomics based in mass spectrometry has been extensively utilized to investigate biological samples for basic and applied research in the clinical field to elucidate changes in the molecular mechanisms caused by diseases. There is still a great lack of information about the molecular mechanisms and the origins of most brain disorders, which makes shotgun proteomics an interesting tool in the study of these diseases. A wide range of samples can be used to study such diseases, such as cerebrospinal fluid, central nervous system cells, and brain tissue via *postmortem* analysis. As such, different protein extraction methods must be applied to achieve the best results for each sample type. The lysis buffer, digestion protocol, and peptide purification steps chosen prior to liquid chromatography-mass spectrometry analyses are essential to obtain reliable proteomic datasets. For this reason, the availability of a list with a variety of methods and a description of the pros and cons for each one has been compiled and elaborated upon. This review presents several methods for protein extraction, protein digestion, and sample cleanup with a focus on shotgun proteomics via mass spectrometry and a further focus on studying brain disorders.

Keywords Sample preparation · Shotgun LC-MS analysis · Brain disorders · Biological samples

---

V. C. Carregari (✉)
Laboratory of Neuroproteomics, Department of Biochemistry and Tissue Biology, Institute of Biology, University of Campinas (UNICAMP), Campinas, Brazil

V. Corasolla Carregari (ed.), *Understanding PTMs in Neurodegenerative Diseases*, Advances in Experimental Medicine and Biology 1382,
https://doi.org/10.1007/978-3-031-05460-0_1

1.1 Introduction

Over the past decade, mass spectrometry techniques have been extensively applied to discovery-based proteomic approaches; and more recently, the number of studies using these methods have increased for translational medicine, shortening the distance between the research bench and clinical diagnostics [1, 2]. Bottom-up, shotgun proteomics is the most used method for the analysis of biological samples thus far to understand complex physiopathology mechanisms, seek new diseases biomarkers, and develop new diagnostic methods [3–6].

Due to its versatility, this method can be applied to several types of biological samples--such as organs, fresh tissue, paraffin-embedded tissue, cell cultures, and body fluids, including blood, plasma, cerebrospinal fluid, saliva, and semen. To extract the protein content from these very diverse samples, there are also many protocols adapted to the conditions of each source [7]. In bottom-up proteomics, the proteins that are extracted must be digested by enzymes with specific cleavage sites, generating thousands to millions of peptides that create a complex puzzle that must be deconvoluted by the software that matches these fragments to a protein databank with a confidence score. To reduce the spectral complexity and permit identification, these digested peptides pass through a separation method, most commonly a liquid chromatographic separation that is coupled to a mass spectrometer, thereby forming the LC-MS method [8].

The majority of multifactorial diseases, including neurodegenerative diseases, are still lacking in regard to the molecular and biochemical mechanisms behind their origins. A large-scale approach like shotgun proteomics has been proven to be much more suitable for molecular mechanism discovery than other proteomic techniques (immunoprecipitation, antibody-dependent, and multiplex techniques), since there is no need for previous knowledge about the sample contents or the biological system to be analyzed, thereby excluding the need to set molecular targets or a protein panel to achieve answers, instead being able to identify qualitatively and quantitatively thousands of proteins [9–11].

To carry out a successful proteomics experiment with high protein coverage and reliable quantification, high-quality and reproductible protein extraction methods are imperative, which are in turn dependent on an effective cell lysis. Methods for protein extraction can vary based on the origin of a sample, using buffers that contain a detergent such as SDS, Triton-100-X, CHAPS, high salt concentration, or other substances to interact and lyse cell membranes. Detergents and high salt concentrations are not directly compatible with LC-MS approaches, therefore making sample cleaning a very important step in shotgun proteomics. Therefore, to obtain high reproducibility, avoid technical problems or damage to the LC system, prevent a loss of important and unique samples, and increase protein coverage and reliability of the data, sample preparation and digestion must be taken seriously and be correctly performed.

In this work, we discuss different methods and buffers for protein extraction, sample cleaning, and optimization methods for protein digestion, hereby creating a guide for the most appropriate sample preparation method that can be used for central nervous system (CNS) cell cultures, brain tissue, and cerebrospinal fluid.

1.2 Lysis Buffers

Before starting sample preparation, it is vital to define which lysis buffer or buffers are most suitable for your sample; and to do so, it is important to first know what a lysis buffer for proteomics must contain as well as the chemical function of each substance. Cell cultures and their corresponding tissue samples are often very similar in their chemical composition, so protein extraction from neuronal cells and brain tissue can most likely use the same lysis buffer.

There are several basic criteria to be considered when preparing a lysis buffer so as to obtain an effective protein extraction: pH, ionic strength, surfactants, and inhibitors. The pH of the solution is critical, since proteins may precipitate or otherwise become unstable when the pH is too far outside the physiological range. To avoid this issue, a buffer system is usually created, the most commonly used system relying on Tris-HCl. The ionic strength of the solution is also important and is usually created with salts. If a buffer has a salt concentration that is too high, proteins may precipitate or separate unpredictably in polyacrylamide gels. If the salt concentration is too low, hydrophobic proteins will not solubilize well and precipitate out of solution. Next, surfactant agents are used, usually detergents. Surfactants help to solubilize hydrophobic and membrane-interacting proteins in the buffer, and also can help to lyse intracellular membranous structures. Lastly, inhibitors are used, varying depending on the experiment at hand. Tissues and cells in general contain a large number of proteases that are released after membrane disruption; these proteases can digest the proteins extracted in an unpredictable manner, therefore generating problems during LC-MS/MS data interpretation. For this reason, protease inhibitor cocktails are crucial in a lysis buffer for proteomic studies. Many proteases require a divalent metal ion to function, so a sequestering agent is also often used to inhibit protease activity, such as EDTA [12, 13]. In other targeted studies, different inhibitors may also be necessary. In phosphoproteomic assays, for example, a phosphatase inhibitor is necessary to avoid losing phosphorylated residues.

1.3 Buffer Compositions, Pros, and Cons

1.3.1 RIPA Buffer

50 mM Tris-HCl, pH 7.4	0.1% SDS
150 mM NaCl	1 mM EDTA
1% triton X-100	10 mM NaF
0.5% sodium deoxycholate	Protease inhibitor cocktail

RIPA is commonly used for proteomics as it uses a nonionic surfactant that is effective at breaking lipid-lipid and lipid-protein interactions, but normally does not break protein-protein interactions. RIPA is usually highly compatible with subsequent protein quantification methods.

1.3.2 Modified RIPA Buffer

100 mM Tris-HCl, pH 8
150 mM NaCl
0.5% triton X-100
1 mM EDTA
Protease inhibitor cocktail

At times, to reduce the number of wash steps that are necessary without losing too much extraction efficiency, a modified version of RIPA can be used.

1.3.3 SDS Sample Buffer

4% SDS
6% glycerol
150 mM Tris-HCl
0.01% bromophenol blue
5% β-mercaptoethanol

This buffer uses an ionic surfactant, that is very capable at breaking protein-protein interactions, though is not as effective at breaking protein-lipid interactions. This buffer also requires the complete removal of SDS from the sample before LC-MS analyses since it is considered a contaminant in liquid chromatography systems and generates a great number of ions in the mass spectrum.

1.3.4 NP-40 Lysis Buffer

50 mM Tris-HCl, pH 8.0
150 mM NaCl
1% Nonidet P-40
Protease inhibitor cocktail

The composition of NP-40 lysis buffer is very similar to RIPA, but instead uses NP-40 as the nonionic detergent.

1.3.5 HEPES Lysis Buffer

20 mM HEPES, pH 7.5
100 mM NaCl
0.05% triton X-100
1 mM dithiothreitol (DTT)
Protease inhibitor cocktail

This buffer uses HEPES as the pH buffering system, which is more capable at buffering pH than Tris-HCl and PBS.

1.3.6 CHAPS Lysis Buffer

7 M urea
2 M thiourea
4% (w/v) CHAPS
30 mM Tris-HCl, pH 8.5

CHAPS is a zwitterionic detergent that possess the properties of both ionic and nonionic detergents, increasing the protein solubility. However, like SDS, CHAPS is incompatible with LC steps and must be carefully removed from the sample before LC-MS.

1.3.7 Urea Buffer

8 M urea
150 mM Tris-HCl
Protease inhibitor cocktail

In this buffer, no membrane-breaking surfactant is used. It can be utilized when the cellular composition must be preserved, such as in organelle enrichment methods or exosome analyses.

1.3.8 Protease Inhibitor Cocktail

5 mM sodium β-glycerophosphate
0.5 mM sodium orthovanadate
1 mM EDTA
0.5 mM phenylmethylsulfonyl fluoride
10 μg/mL aprotinin
5 μg/mL leupeptin
2 μg/mL pepstatin

When a commercially available protease inhibitor cocktail is not available, a series of metal chelators and molecular inhibitors of proteases can be used instead, as listed above. All concentrations are in the final volume of lysis buffer being used.

1.4 Sample Preparation Methods

To make an informed choice on the lysis buffer to be used, the cell type or tissue must be considered. From there, it must be determined whether a strong surfactant and more robust extraction is necessary or if a gentler extraction can be performed with a low surfactant concentration or even absence thereof. It is always important to remember that higher surfactant concentrations also lead to more steps during sample cleanup, which always have a varying but inescapable loss in protein yield. Once the buffer is chosen, the protein extraction procedure can be executed. In this review, we will explain sample preparation in a step-by-step manner.

1.4.1 Central Nervous System (CNS) Cells

Studies that involve neurological conditions and the central nervous system in general are highly dependent on cellular models due to the low availability of *in vivo* samples. While biofluids are important targets of study for neuropsychiatric disorders and neurodegenerative diseases, particularly in the area of biomarkers, the blood-brain barrier impedes the transfer of the vast majority of secreted macromolecules, including proteins. Due to this separation, proteomic abnormalities are not possible to be investigated using peripheral cells.

The majority of the membranes of CNS cells is composed of phospholipids and sphingolipids. Moreover, these cells do not export high levels of extracellular matrix proteins, which are generally highly hydrophobic. As such, buffers containing non-ionic surfactants are highly suitable for this cell type.

First, if the cells are adherent, they must be detached from the cell culture plate. This can be done using trypsin before centrifuging the cells for 10 min at 1000x*g* to pellet the cells. The culture medium should then be removed, and the pellet should be lightly washed with PBS before repeating the centrifugation step. Otherwise, if the cells are suspended in the medium, the trypsinization step is not necessary. In either case, the centrifugation-PBS-centrifugation steps should be followed. In either case, remove as much PBS as possible to not dilute the lysis buffer.

- Add lysis buffer to the cell pellet at approximately a 2:1 ratio (v/v) and leave the cells on ice to incubate for 30 min.
- Several methods for mechanical cell lysis exist; however, for this cell type, we recommend ultrasonication via probe. To do so, applying 60% of duty cycle during 3 cycles of 20 s each.

- Centrifuge at 1000x*g* for 10 min at 4 °C to pellet the cell debris for 10 min. Remove the protein-containing supernatant and transfer it to a clean tube.
- Quantify the protein content in the lysate. BCA and Bradford are the most frequently used methods for shotgun proteomics. If another method is used, always ensure that the lysis buffer reagents are compatible with the quantification method.

1.4.2 Postmortem Brain Tissue

Brain tissue sample is composed of CNS cells, connective tissue coming from the pia mater, and the extracellular matrix, which is a complex macromolecular network made of proteins and polysaccharides. Due to the complexity of the composition of this tissue, the protein extraction method is a little bit different when compared to cultured cells.

- For each brain slice of 10 mm, add 100 μL of lysis buffer. Ensure that the tissue is entirely covered with buffer.
- Macerate the tissue with plastic pestles until the tissue appears to be 100% homogenized. Ensure that the number of maceration motions is the same for each sample to ensure the highest replicability.
- Incubate the macerated samples in lysis buffer on ice for approximately 30 min.
- To better homogenize the solution and extract proteins, perform an ultrasonication step. Using a probe, sonicate the samples for 3 cycles of 30 s at 90% duty cycle.
- Centrifuge at 1000x*g* for 10 min at 4 °C to pellet the cellular debris.
- Transfer the protein-containing supernatant to a new tube. If desired, store the cell pellet at −80 °C for lipidomics or other analyses.
- Use the supernatant to quantify the protein amount. BCA and Bradford are the most frequently used methods for shotgun proteomics. If another method is used, ensure that the lysis buffer reagents are compatible with the quantification method.

1.4.3 Cerebrospinal Fluid (CSF)

Cerebrospinal fluid (CSF) is an important source of proteins for discovery-based proteomics since it flows throughout the CNS and is in direct contact with CNS cells and contains their secreted proteins. As such, it is able to reflect changes that are induced by a wide range of brain disorders. Proteins in the CSF, however, are very diluted in the liquid medium, additionally containing a large amount of carbohydrates [14]. To precipitate proteins from the rest of the medium, we present two recommended methods.

Methanol-based precipitation:

- For every 100 μL of CSF, add 400 μL of 100% methanol. For most proteomic experiments, approximately 200 μL of CSF are recommended.

TCA-acetone-based precipitation:

- For every 100 μL of CSF, add 400 μL of precipitation buffer, composed of 10% trichloroacetic acid (TCA) in acetone containing 2% β-mercaptoethanol.
- Mix by inverting the tube 15 times and let the proteins precipitate for 12 h at −20 °C.
- Centrifuge at 5000x*g* for 30 min, then wash the pellet three times with 1 mL ice-cold acetone, vortexing the Eppendorf to homogenize the pellet. Centrifuge at 5000x*g* for 30 min after each wash step to re-pellet the proteins.
- Let the acetone evaporate at room temperature, then freeze the pellets at −80 °C until use.

If metabolomic assays are also to be carried out with the CSF samples, the following method allows the separation of the hydrophilic (polar) phase from the hydrophobic (nonpolar) phase.

Two-phase precipitation:

- Add approximately 200 μL of CSF to a 3 k Amicon column (Merck) and complete to 1 mL with PBS.
- Centrifuge at 4500x*g* for 45 min at 4 °C. Add PBS to the column to complete to approximately 1 mL and repeat the centrifuge step. Repeat this step another 3 times to perform a total of five centrifugations.
- After each centrifugation, remove the flowthrough. If performing a metabolomic analysis, save the solution in a Falcon tube at −80 °C until use. Otherwise, discard the flowthrough.
- Transfer any remaining solution in the Amicon column to a fresh tube and extract any membrane-bound proteins by washing the filter membrane with a 2 M urea solution and adding the wash to the same tube.
- The extracted proteins can now be quantified.

1.5 Digestion Methods for Shotgun Proteomics

Bottom-up proteomics is extremely dependent on a complete and efficient protein digestion step into constituent peptides. Data quality and the depth of the coverage of the proteome are strictly related to and dependent on the chosen sample preparation method and separation technique or techniques.

Though other methods exist, we will cover three recommended digestion protocols: in solution digestion, C18 micro-column digestion, and non-resin filter-based digestion (divided into FASP and S-TRAP, both described in detail below). In the column-based digestion protocols, the sample is cleaned after the trypsin digestion by either a C18 micro-column or other filter used to remove the lysis buffer, exchanging it to a buffer more suitable for enzymatic digestion. As previously mentioned, detergents and high quantities of salt are not compatible with the LC-MS system, these compounds must be removed if present for a quality result [15, 16].

1.5.1 In-Solution Digestion

In this protocol, proteins are digested directly in solution, using a dilution buffer to increase sample solubility and decrease detergent interference. After trypsin digestion, all peptides must go through a C18-reverse phase chromatographic step for cleanup and to remove excess urea and any detergent.

- After protein quantification, add 50 μg of the protein extract and add the same volume of dilution buffer to the sample (Tris-HCl 50 mM, urea 6 M).
- Protein digestion can be impeded by disulfide bridges. To break these bonds, reduce the proteins with DTT at a final concentration of 15 mM. Incubate the samples for 1 h in a thermomixer set to 500 rpm and 56 °C.
- To ensure the disulfide bridge does not spontaneously reform, proteins must be alkylated with iodoacetamide (IAA) at a final concentration of 50 mM. Incubate the samples for 30 min at room temperature. Since IAA is light-sensitive, perform this incubation in the dark. Note that this step transfers a carbamidomethyl group to cysteine residues, which should be selected as a fixed modification during downstream protein identification steps.
- To quench any excess of IAA and ensure that it does not impede proper trypsin function, add DTT to a final concentration of 8 mM and incubate the samples with gentle agitation for 30 min at 56 °C in a thermomixer system. Cover with tinfoil.
- Before adding trypsin, the urea must be diluted to a final concentration of 2 M from its original concentration of 6 M.
- Add 1 μg of trypsin to each sample (trypsin:protein ratio of 1:50) and incubate at 37 °C and 300 rpm. Digest overnight, between 12 and 18 h, covered with tinfoil.
- The next day, quench trypsin activity by decreasing the pH; add formic acid (FA) to a final concentration of 1% (v/v).
- Dry the digested peptides in a speed vacuum system for LC-MS/MS analysis.
- To remove any remaining detergent and salt, resuspend the dried peptides in 50 μL of 0.1% FA and then pass through a C18 reverse-phase microcolumn, as described below.

1.5.2 Homemade C18 Microcolumn

Gobom et al. originally developed a purification method that use a chromatographic resin packed in a small, constricted tip, forming a microcolumn [17]. The microcolumns can be prepared in any pipette tip, and in this example, we will show how to create a microcolumn with a 200 μL tip.

- Cut a 200 μL (P200) tip to fit a plastic syringe (Fig. 1.1, Step 1).
- Cut a small piece of a 3 M Empore C18 disk in a way that it can be compressed in the end of the tip (Fig. 1.1, Step 2).
- Dilute Oligo R3 material in 100% acetonitrile, then transfer 40 μL of this R3 slurry to the tip. (Fig. 1.1, Step 3).
- Connect the syringe to the tip and gently apply positive pressure to expel the ACN, packing the column (Fig. 1.1, Step 4).
- Equilibrate the column with 40 μL 0.1% formic acid (FA), using the same air pressure method. Do not completely dry the column here.
- Load 10 μL of peptides (maximum 5 μg of digested peptides) diluted in 0.1% FA onto the column, topping it off with another 30 μL of 0.1% FA. Apply pressure with the syringe without letting the resin run dry.
- Wash the column with 40 μL 0.1% FA and apply pressure to completely dry the column.
- Into a clean tube, elute the bound peptides first with 40 μL of 60% ACN/0.1% FA. Then completely wash any remaining peptides out with 40 μL of 80% ACN/0.1% FA.
- Dry the eluted peptides in a speed vacuum system.

One of the advantages of this method is the possibility to digest higher protein amounts then other filter-based digestion methods. However, a considerable drawback is the time and labor necessary to carry out this protocol. Moreover, the

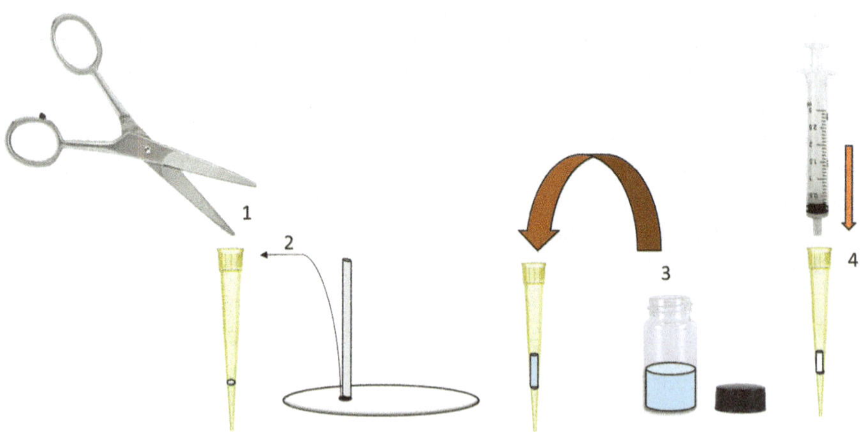

Fig. 1.1 Diagram depicting a basic outline of the steps necessary to create a P200 microcolumn

chromatographic step can result in lost peptides, potentially decreasing reproducibility, or can result in residual detergent in the sample.

1.5.3 Filter-Aided Sample Preparation (FASP)

This protocol uses an 8 M urea buffer to disrupt detergent micelles in combination with with a filter (3 kDa, 10 kDa, or 30 kDa) before washing samples several times to completely remove the detergent and exchange the urea buffer for 50 mM ammonium bicarbonate (AmBic) for trypsin digestion [18]. An adapted FASP protocol is described below, based on modifications by Marini et al. [19].

- If quantitative analyses will be carried out, use protein quantification to normalize all the samples for total protein amount.
- Rinse the centrifugal filter by adding 100 μL of 1% (vol/vol) FA to the filter unit. Centrifuge for 15 min at 14,000x*g*.
- Transfer between 10 and 100 μg of protein extract to the centrifugal filter and complete to 200 μL with Urea Buffer (Tris-HCl 100 mM, 8 M urea, pH 8). Centrifuge the mixture for 20 min.
- Wash the samples by adding 200 μL of Urea Buffer to each filter unit. Centrifuge for 20 min at 14,000x*g*.
  - If the lysis buffer contained high SDS concentration, repeat this washing step two times, for a total of 3 washes.
- Reduce the proteins with 100 μL DTT solution (8 mM). Incubate, covered with tinfoil, with gentle agitation for 40 min at 56 °C in a thermomixer system.
- Centrifuge at 14,000x*g* for 15 min.
- Add 100 μL of Urea Buffer to the centrifugal filter to wash any residual DTT. Centrifuge for 15 min at 14,000x*g*.
- Alkylate the proteins by adding 100 μL of iodoacetamide (IAA; 50 mM). Incubate in the dark for 20 min at room temperature.
- Centrifuge at 14,000x*g* for 15 min.
- Wash the remaining IAA with 100 μL of Urea Buffer. Centrifuge for 15 min at 14,000x*g*.
- Quench any excess IAA by adding 100 μL of DTT solution. Incubate with gentle agitation for 15 min at 56 °C in a thermomixer system, covered with tinfoil.
- Centrifuge at 14,000x*g* for 15 min.
- To wash the remaining DTT, add 100 μL of Urea Buffer to the centrifugal filter and centrifuge for 45 min at 14,000x*g*.
  - Ensure that all buffer has passed through the filter to ensure a complete buffer exchange with AmBic during the following steps.
- Add 100 μL of AmBic (50 mM) and centrifuge at 14,000x*g* for 15 min. Repeat this step twice for a total of three additions.

- Move the column to a new, low-binding collection tube.
- Add approximately 100 μL of trypsin working solution to the filter and incubate at 37 °C with 300 rpm agitation to digest overnight in thermomixer system. Seal the tubes with parafilm to prevent the sample from evaporating, drying the filter.
  - Ensure that the trypsin:protein ratio is 1:50, adjusting the concentration of trypsin based on how much protein was applied to each column. Ensure that at least 40 μL are used, otherwise the filter may not be completely saturated and dry spots will remain.
- After overnight digestion, centrifuge for 15 min at 14,000x*g* and collect the peptide-containing flowthrough.
- Add 50 μL of AmBic (50 mM) to the centrifugal filter and centrifuge for 15 min at 14,000x*g*, collecting the flowthrough in the same tube.
- Acidify the flowthrough with FA to reach 1% final concentration (vol/vol).

This method enables sample cleanup, reduction, alkylation, and digestion all in the same column, increasing reproducibility and downstream quantitative analyses. The time necessary to carry out this protocol is reduced compared to in-solution digestion, and the efficiency of detergent removal is also much higher.

1.5.4 Suspension Trap (S-TRAP)

Differently to the conventional filter-aided sample preparation (FASP) approach, S-Trap does not utilize a molecular weight cut-off (MWCO) membrane as a filter, thereby decreasing manyfold the time required for sample processing until digested peptides are obtainable. S-Trap employs a three-dimensional porous material that works as a filter medium and traps all proteins in suspension, depleting the interfering substances before subsequently performing in-filter digestion [20].

- Add 23 μL of the chosen lysis buffer to pelleted the cells .
  - For good recovery, the SDS concentration must be higher than 2%. If the chosen lysis buffer does not already contain it, add SDS until 2% is reached.
  - S-Trap is not affected by high concentrations of SDS (up to 15% is acceptable). If necessary, samples can be concentrated in a vacuum centrifuge.
- Vortex to lyse the sample and suspend proteins. If the sample is particularly viscous, it is probably due to the presence of DNA presence. Sonicate the sample or treat with bezonase as DNA can clog the S-Trap column.
- Centrifuge the mixture at 13000x*g* for 8 min.
- Quantify the protein amount before to follow to the next step. If you are willing to perform further quantification analysis, normalize the protein amount for each samples before to load into the S-Trap column.

- Reduce the proteins by adding 1 μL (5 mM final concentration) of Tris(2-carboxyethyl)phosphine (TCEP) and incubate at 55 °C for 15 min.
- Alkylate the disulfide bonds with 1 μL (20 mM final concentration) of methyl methanethiosulfonate (MMTS) and incubate at room temperature for 10 min.
- Add 2.5 μL of phosphoric acid, 27.5% in water, to reach a final concentration of approximately 2.5% phosphoric acid. Vortex well. This step is crucial to trap proteins efficiently.
- Add 165 μL of Wash Buffer (TEAB, final concentration 100 mM, in 90% methanol) to the sample and mix. Place the S-Trap column in a 1.5 mL tube and transfer all the sample volume to the column.
- Centrifuge for 60 s at 4000x*g* to trap the proteins.
- Add 150 μL of Wash Buffer and centrifuge for 60 s at 4000x*g*. Repeat this step 3 times for a total of 4 washes, discarding the flowthrough as necessary.
  - Usually, brain tissue has particularly hydrophobic contaminants such as lipids. In this case, wash the samples three times with 50% $CHCl_3$/50% MeOH, filling the entire column each time, then perform three washes with standard Wash Buffer.
- The Wash Buffer must be entirely removed, so centrifuge the column for 90 s at 4000x*g*. If needed, increase the centrifugation time.
- Transfer the column to a new, low-binding tube.
- Add 20 μL of 50 mM AmBic containing trypsin, following a 1:10 trypsin:protein (wt:wt) ratio to the S-Trap.
  - For effective digestion, use at least 1 μg of trypsin, regardless of the protein weight used.
- Cap the S-Trap loosely to limit evaporative loss and let the samples incubate overnight at 37 °C in a thermomixer.
- Elute the peptides with 40 μL of 50 mM triethylammonium bicarbonate, pH 8.5 (TEAB), and centrifuge at 4000x*g* for 1 min.
- Add 40 μL of 0.2% formic acid in water and repeat the centrifugation step.
- To recover more hydrophobic peptides, add 40 μL of 50% acetonitrile in water to the column and centrifuge at 4000x*g* for 1 min.
- Dry the eluted peptides in a speed vacuum system.

The biggest advantage of this method is a 3 h reduction in the total processing time compared to the FASP protocol while still presenting a great effectiveness at contaminant removal. Similar to the FASP protocol, however, one of the limitations of this method is the limited amount of protein that can be used for digestion. The efficiency of each protocol with comparisons on their advantages and disadvantages of each method have already been published [15, 20].

1.6 Conclusions

Shotgun proteomics has a fundamental place in the clinic, able to be applied from discovery to basic science, also including applied science and diagnostical analyses. The success of an assay and the quality of the data obtained from it are tightly related to the care taken when preparing samples, which includes cell lysis and protein extraction and purification. These critical steps ensure high reproducibility, and consequently high robustness and reliable quantification. This review listed some of the pros and cons of different methods with a focus on samples obtained while studying neurological and CNS disorders, providing readers with information to make better choices when designing experiments. It is important to always keep in mind that combining multiple methods in a single experiment can sometimes lead to the best possible results.

References

1. Walther, T. C., & Mann, M. (2010). Mass spectrometry-based proteomics in cell biology. *The Journal of Cell Biology, 190*, 491–500.
2. Paik, Y.-K., et al. (2019). Advances in identifying and characterizing the human proteome. *Journal of Proteome Research, 18*, 4079–4084.
3. Hood, L., & Perlmutter, R. M. (2004). The impact of systems approaches on biological problems in drug discovery. *Nature Biotechnology, 22*, 1215–1217.
4. Köcher, T., & Superti-Furga, G. (2007). Mass spectrometry–based functional proteomics: From molecular machines to protein networks. *Nature Methods, 4*, 807–815.
5. Gstaiger, M., & Aebersold, R. (2009). Applying mass spectrometry-based proteomics to genetics, genomics and network biology. *Nature Reviews. Genetics, 10*, 617–627.
6. Han, X., Aslanian, A., & Yates, J. R. (2008). Mass spectrometry for proteomics. *Current Opinion in Chemical Biology, 12*, 483–490.
7. Kwasnik, A., et al. (2016). Proteomes, their compositions and their sources. *Advances in Experimental Medicine and Biology, 919*, 3–21.
8. Syka, J. E. P., Coon, J. J., Schroeder, M. J., Shabanowitz, J., & Hunt, D. F. (2004). Peptide and protein sequence analysis by electron transfer dissociation mass spectrometry. *Proceedings of the National Academy of Sciences, 101*, 9528–9533.
9. Wolters, D. A., Washburn, M. P., & Yates, J. R. (2001). An automated multidimensional protein identification technology for shotgun proteomics. *Analytical Chemistry, 73*, 5683–5690.
10. Obrocki, P., et al. (2020). Perspectives in fluid biomarkers in neurodegeneration from the 2019 biomarkers in neurodegenerative diseases course—A joint PhD student course at University College London and University of Gothenburg. *Alzheimer's Research & Therapy, 12*, 20.
11. Corasolla Carregari, V., et al. (2020). Proteomics of muscle microdialysates identifies potential circulating biomarkers in Facioscapulohumeral muscular dystrophy. *International Journal of Molecular Sciences, 22*.
12. Peach, M., Marsh, N., & MacPhee, D. J. (2012). Protein Solubilization: Attend to the choice of lysis buffer. In B. T. Kurien & R. H. Scofield (Eds.), *Protein electrophoresis: Methods and protocols* (pp. 37–47). Humana Press. https://doi.org/10.1007/978-1-61779-821-4_4
13. Peach, M., Marsh, N., Miskiewicz, E. I., & MacPhee, D. J. (2015). Solubilization of Proteins: The Importance of Lysis Buffer Choice. In B. T. Kurien & R. H. Scofield (Eds.), *Western Blotting: Methods and Protocols* (pp. 49–60). Springer. https://doi.org/10.1007/978-1-4939-2694-7_8

14. Hörmann, P., Barkovits, K., Marcus, K., & Hiller, K. (2019). Co-extraction for metabolomics and proteomics from a single CSF sample. *Methods in Molecular Biology (Clifton, N.J.), 2044*, 337–342.
15. Ludwig, K. R., Schroll, M. M., & Hummon, A. B. (2018). Comparison of in-solution, FASP, and S-trap based digestion methods for bottom-up proteomic studies. *Journal of Proteome Research, 17*, 2480–2490.
16. Elinger, D., Gabashvili, A., & Levin, Y. (2019). Suspension trapping (S-trap) is compatible with typical protein extraction buffers and detergents for bottom-up proteomics. *Journal of Proteome Research, 18*, 1441–1445.
17. Gobom, J., Nordhoff, E., Mirgorodskaya, E., Ekman, R., & Roepstorff, P. (1999). Sample purification and preparation technique based on nano-scale reversed-phase columns for the sensitive analysis of complex peptide mixtures by matrix-assisted laser desorption/ionization mass spectrometry. *Journal of Mass Spectrometry, 34*, 105–116.
18. Wiśniewski, J. R., Zougman, A., Nagaraj, N., & Mann, M. (2009). Universal sample preparation method for proteome analysis. *Nature Methods, 6*, 359–362.
19. Marini, F., et al. (2020). Exploring the HeLa dark mitochondrial proteome. *Frontiers in Cell and Development Biology, 8*, 137.
20. HaileMariam, M., et al. (2018). S-trap, an ultrafast sample-preparation approach for shotgun proteomics. *Journal of Proteome Research, 17*, 2917–2924.

Chapter 2
Phosphopeptide Enrichment Techniques: A Pivotal Step for Phosphoproteomic Studies

Victor Corasolla Carregari

Abstract Many pathological conditions are caused by dysregulation of cell signaling, which can generate a cascade of abnormal responses and completely change the functions of a cell or tissue. A large portion of the regulation of these signals is via protein phosphorylation, in which cell responses can be activated or inhibited. Proteins that are both downstream and upstream of a phosphorylated protein can be modified, altering metabolism and other biological processes. Recently, the number of phosphoproteomic studies based on mass spectrometry has increased, constantly aiming to obtain a higher coverage of proteins and increase the number and location of their phospho-sites, as well as better understand their respective phosphorylation states. In this way, it is possible to better understand biological processes as a whole and their roles in cellular dysfunctions and diseases. To study changes at the phosphoproteome level, the stochiometric imbalance between the non-phosphorylated and phosphorylated peptides must be overcome, since higher quantities and comparatively better ionization of non-phosphorylated peptides can suppress the ion signals of the phosphorylated peptides. It is for this reason that phosphophopeptides are rarely found in samples that did not pass through a phospho-enrichment step, highlighting the importance of performing enrichment steps in phosphoproteomic studies. The numbers of identified phosphopeptides and phosphorylation sites are extremely important to the quality of a phosphoproteomic analysis; therefore, the efficiency of the enrichment process is critical. Here, phospho-enrichment techniques are presented to offer insight into the applicability of these methods to different experiment types and consequently support the growth of phosphoproteomic studies overall.

Keywords Phosphoproteomics · Mass spectrometry · Phospho-enrichment · Phosphopeptides

V. C. Carregari (✉)
Laboratory of Neuroproteomics, Department of Biochemistry and Tissue Biology, Institute of Biology, University of Campinas (UNICAMP), Campinas, Brazil

V. Corasolla Carregari (ed.), *Understanding PTMs in Neurodegenerative Diseases*, Advances in Experimental Medicine and Biology 1382,
https://doi.org/10.1007/978-3-031-05460-0_2

2.1 Introduction

Mass spectrometry-based proteomics is a powerful tool to understand the regulation of the proteome in tissues, cells, and biofluids in conditions such as after a diet or other treatment and in diseases or disorders, revealing altered biological mechanisms and possible targets for biomarkers and novel treatments. Despite the many benefits of proteomics, several bioprocesses such as signal transduction, molecular recognition and interaction, response regulation of external stimuli, and many other cellular events are additionally regulated by protein phosphorylation and dephosphorylation [1, 2], details of the proteome that are not obtainable with standard shotgun methods. Almost 30% of proteins in mammalian cells at some point during their expression are thought to be phosphorylated [3], making phosphorylation one of the most widespread post-translational modifications (PTMs).

A better comprehension of biological processes that can be activated or inhibited by the phosphorylation state of a protein and respective upstream and downstream effects is fundamental to unravel important mechanisms that can be (dys)regulated in a pathological condition. Despite the fact that almost 1/3 of all proteins are potentially phosphorylated, a phosphoproteome exhibits a significant increase in complexity when compared to a normal proteome since many of these modified proteins that involved in signaling and the regulation of cellular functions are less abundant than structural and cytosolic proteins [4, 5].

When protein phosphorylation occurs, a phosphate group is transferred to a serine (S), threonine (T), or tyrosine (Y) residue, which is then denominated as a phospho-site [6]. Phosphoproteins usually contain between 1 and 20 phospho-sites, and these sites are largely be occupied at sub-stoichiometric levels [7]. Further increasing proteome complexity, proteins can also have many other PTM sites, many of which can also co-occur at phosphorylation sites. Moreover, since tryptic peptides can present multiple phosphorylation sites, the *in silico* analysis of these data becomes even more challenging. Overall, these factors are demonstrated by the measured complexity of the phosphoproteome, which adds approximately 2 orders of magnitude in the dynamic range compared to the general proteome, since phosphopeptides exist at much lower levels than their non-phosphorylated counterparts [8].

The first step when studying phosphoproteins is to carefully plan and carry out sample preparation. After cell lysis, many active proteases and phosphatases are released, degrading proteins and cleaving phosphate groups in the protein extract if left unchecked. Working at a low temperature (4 °C) reduces protease activity, but low temperatures cannot be feasibly maintained during the protein digestion step, in which an enzyme such as trypsin is used [9]. While most lysis buffers include protease inhibitor cocktails, protecting the sample from protease activity, phosphatases can still act upon the lysate, thereby decreasing phosphoproteome coverage and negatively affecting comparisons. To overcome this, it is important to choose a correct mix of phosphatase inhibitors to be added to the lysis buffer before performing cell lysis so as to maintain the biological phosphorylation states of the proteome [10].

2.2 Phosphopeptide Enrichment

To overcome the stoichiometric imbalance of phosphopeptides, the application of sensitive and selective methods capable of enriching/capturing these peptides based on their modification state is essential. Phosphopeptide enrichment is mandatory to perform phosphoproteomic analyses, since normal shotgun approaches are rarely able to sufficiently detect these peptides due to an ion suppression effect from the abundant unmodified peptides [6, 11]. The high abundance, unphosphorylated proteins must then be removed from the sample before analysis to ensure a complete acquisition of phosphopeptide data.

Several methods for phosphopeptide enrichment have been developed, many of which use metal ions to reversibly bind phosphate groups, thereby selecting only the phosphopeptides while unmodified peptides are washed away (Fig. 2.1). Immobilized Metal Affinity Chromatography (IMAC) is one of the most popular methods [10, 12]. Other methods such as immuno-affinity based methods [13] and titanium dioxide (TiO_2) chromatography [14] are also extensively used in the field. In some cases, a combination of methods is also possible to be employed to further increase phosphoproteome coverage. Despite the efforts made to develop new and more efficient methods for phosphopeptide enrichment, it is important to understand that no

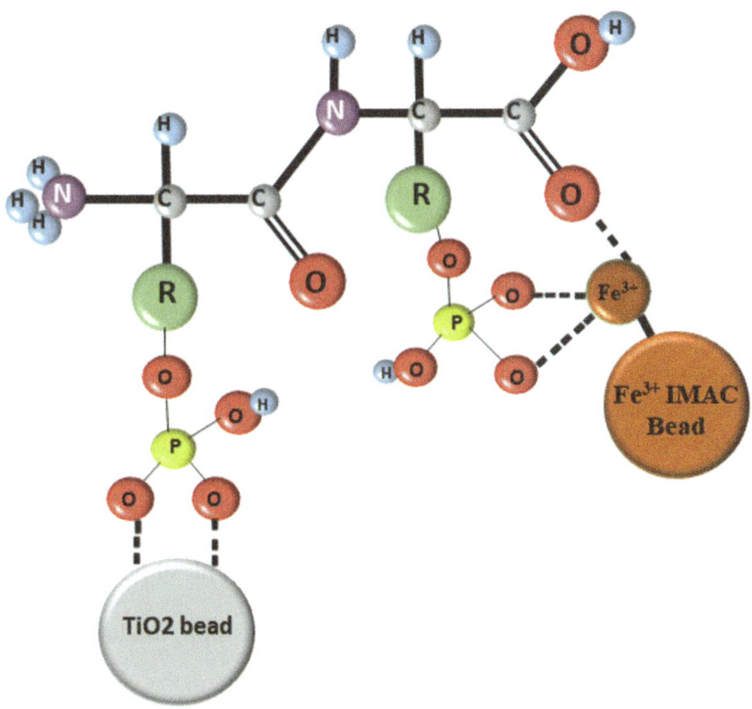

Fig. 2.1 Example of the interactions between phosphopeptides and IMAC and TiO2 beads

currently known method or combination thereof is able to offer a complete coverage of the phosphoproteome [15]. In this work, protocols and variations for two of the most used phosphoenrichment techniques for mass spectrometry-based phosphoproteomic analyses, IMAC and TiO_2, are briefly explained.

2.3 IMAC

IMAC, in brief, is carried out by allowing phosphorylated peptides to reversibly bind to a chelating material containing metal ions, which interacts directly with the phosphate group present on the phosphorylated amino acid residue [16–18]. Iron and gallium are the most commonly used metal ions for this application. In this method, peptides must be present in a solution containing a weak acid to equilibrate the protonation state of the molecules and enable subsequent binding to the IMAC material. The interactions between the IMAC material and the phosphopeptides are sensitive to pH change, and metal chelators and other certain reagents can interfere with the interactions between the metal ion and the phosphate group, including phosphoric acid/phosphate buffers. Using these chemical characteristics, an elution buffer can be developed to elute bound peptides after non-interacting peptides are washed away.

1. Calculate the amount of IMAC beads that will be used for the enrichment. 500 μg of a digested peptide mixture is recommended, for which 10 μL of IMAC beads will be needed (PhosSelect, Sigma). It is important to use a correct ratio of digested peptides to IMAC beads to avoid non-specific binding by non-phosphorylated peptides.
2. Transfer 10 μL of IMAC beads to a 1.5-mL low-binding tube. Wash the beads[1] with 50 μL of loading buffer (50% acetonitrile in water, with 0.1% trifluoracetic acid, v/v).
3. Dilute the beads in 50 μL of loading buffer and add 500 μg of the digested peptides.
4. Incubate the mixture in a thermomixer at 500 rpm at room temperature.
5. Compress the end of a GELoader pipette tip (Eppendorf) until the IMAC beads do not pass through it. Apply positive pressure with a syringe at the top of the tip to pass the liquid through the compressed tip so that only the IMAC beads remain, thereby generating an IMAC micro-column.
6. Wash the IMAC column once using 50 μL of loading buffer.
7. Add 30 μL of elution buffer (1% ammonia in water, pH 11; this buffer can be more easily made by adding 960 μL of water to 40 μL of a 25% ammonia in water solution).
8. Using the syringe, apply pressure slowly, drop by drop, to elute the peptides. Repeat this step once.

[1]After this and all future wash steps, centrifuge the sample at 1000x*g* for 1 min and discard the supernatant.

9. Dry the phosphopeptides in a vacuum centrifuge and store at −80 °C until mass spectrometry analysis.

2.4 TiO_2

TiO_2 is an alternative method to IMAC, first described in 2004 [14, 19]. The first studies to use TiO_2 enrichment exhibited a high binding of non-phosphorylated peptides, bringing up the necessity to optimize the initial peptide loading buffer and introduce more rigid buffer conditions to load the peptides onto the TiO_2 resin. The substitution of aromatic acids and the use of a lower pH allowed for a reduction in non-specific binding [14, 20, 21].

1. Digest 600 μg of protein extract with trypsin and dilute in 1 mL of loading buffer (80% acetonitrile [ACN], 5% trifluoroacetic acid [TFA], 1 M glycolic acid).
2. For each 100 μg of peptides to be enriched, add 0.6 mg of TiO_2 beads to the peptide solution.
3. Place the tubes in a thermomixer system at 1000 rpm at room temperature for 15 min.
4. Centrifuge for 1 min at 1000x*g* at room temperature to spin down the beads and remove the supernatant, placing it into a new tube.
5. Wash the beads with 100 μl of loading buffer and vortex it briefly.
6. Centrifuge as in step 4 to spin down the beads again and remove the supernatant, placing it into the same tube as before.
7. To increase phosphopeptide recovery, transfer the supernatant pool from steps 4 and 6 to a new batch of TiO_2 beads and repeat the incubation and wash steps.
8. Wash the two combined batches of TiO_2 beads with 100 μl of loading buffer and centrifuge at 1000x*g* for 1 min. Discard the supernatant.
9. Wash the beads with 100 μL of 80% ACN, 1% TFA. Vortex briefly and centrifuge for 1 min at 1000x*g* to pellet the beads.
10. Discard the supernatant and wash the beads with 20% ACN, 0.5% TFA. Centrifuge to spin down the beads and discard the supernatant.
11. Dry the beads in a speed vacuum system for 5 min.
12. Elute the phosphopeptides with 50 μL elution buffer (5% ammonium hydroxide, pH 11.3) and vortexing for 15 min.
13. Centrifuge for 1000x*g* for 1 min and deplete the TiO_2 beads. Phosphopeptides are diluted in the supernatant. Pass the supernatant through a filter (C8 stage tip) and recover the liquid in a 1.5-mL low-binding tube.
14. Wash the beads with 50 μL elution buffer, centrifuge to pellet the beads, and pass the supernatant through the C8 filter into the same low-binding tube.
15. Dry the phosphopeptides in a vacuum centrifuge and store at −80 °C until mass spectrometry analysis.
16. If desired, the washed and otherwise discarded solutions can be pooled and dried to perform a non-phosphopeptide analysis and/or evaluate the phosphoenrichment efficiency.

2.5 TiO_2 Micro-Column

To reduce the time necessary for the phosphoenrichment protocol, it is possible to create a TiO_2 micro-column, speeding up centrifugation steps (Fig. 2.2).

1. Create a TiO_2 micro-column using a GELoader tip. Alternatively, these micro-columns can be purchased pre-made (GL Sciences, Titan sphere TiO_2, 5 microns).
2. Connect a centrifugal adapter to a waste fluid tube. Insert the micro-column into the tube.
3. Wash the column with 20 μL buffer A (1% TFA in 80% ACN). Centrifuge at 3000x*g* for 2 min.
4. Wash the column with 20 μL of buffer B (80% ACN, 5% TFA, 1 M glycolic acid). Centrifuge 3000x*g* for 2 min. Discard the 40 μL of flow-through.
5. Add between 15 μL and 50 μL of sample along with 100 μL of buffer B to the micro-column. Centrifuge at 1000x*g* for 10 min.
6. Repeat the centrifuge step with another 100 μL of buffer B. Discard the flow-through.
7. Add 20 μL of buffer A to the micro-column. Centrifuge at 3000x*g* for 2 min. Add 20 μL of buffer B to the micro-column and centrifuge at 3000x*g* for 2 min. Discard the flow-through.

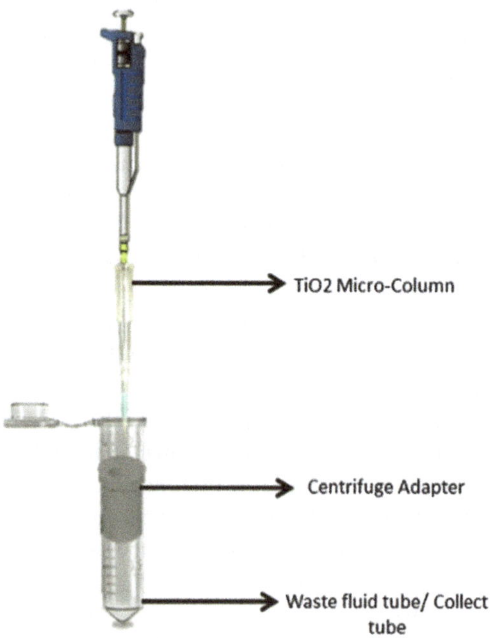

Fig. 2.2 Schematic of a TiO2 micro-column

8. Replace the waste tube with a clean, low-binding tube. Add 50 μL of elution buffer (5% ammonium hydroxide, pH 11.3) and centrifuge at 1000x*g* for 5 min. Repeat this step with another 50 μL of elution buffer and centrifuge run twice more.
9. Dry the resulting phosphopeptides in a vacuum centrifuge and store at −80 °C until analysis by mass spectrometry.

2.6 SIMAC

Sequential elution from IMAC (SIMAC) is a method that was originally developed to gain sensibility in the separation of multiply phosphorylated peptides, which is a recurring challenge in large-scale phosphoproteomic studies. TiO_2 chromatography exhibits a preference for mono-phosphorylated peptides, reducing the number of identifiable, multiply phosphorylated peptides via TiO_2 chromatography. This is due to the high affinity of TiO_2 beads to these multiply phosphorylated peptides, making it much more difficult to disrupt this binding with an elution buffer; as a result, singly phosphorylated proteins are most readily eluted and subsequently analyzed [10]. Contrarily, IMAC seems to work best for multiply phosphorylated peptides [11, 21].

As with the comparison between modified and unmodified peptides, the higher concentrations of mono-phosphorylated peptides and better ionization compared to multiply phosphorylated peptides causes an ion suppression during spectrum acquisition. These factors provide an explanation as to why mono-phosphorylated peptides are predominantly identified in phosphoproteomic studies. Thingholm et al. [10] developed and published a new method that combines IMAC with TiO_2 chromatography, named SIMAC (Sequential elution from IMAC). In this protocol, singly and multiply phosphorylated peptides are enriched from complex peptide mixtures into separate fractions, allowing MS/MS analyses of both types of phosphorylated peptide ions.

SIMAC allows for an increase in the number of identifications of multiply phosphorylated peptides by almost twofold. Moreover, the separation of mono-phosphorylated peptides from multiply phosphorylated peptides also increases the identification of the mono-phosphorylated peptides since the phosphorylated peptides no longer compete to bind to the beads, improving the interactions between mono-phosphopeptides and TiO_2 beads [22]. SIMAC has proven to be a rapid and relatively simple method to improve the number of identified phosphopeptides, making it a great advance in large-scale phosphoproteomics.

1. Measure 300 μg of peptides digested with trypsin in digestion buffer.
2. Wash 50 μL of IMAC beads (PHOS-select Sigma) twice with loading buffer A (50% ACN in water, with 0.1% TFA v/v). Perform a brief centrifugation after each wash and discard the supernatant, leaving the beads.

3. Add 50 µL of loading buffer A to the IMAC beads and transfer to the digested peptide solution. Incubate in a thermomixer at room temperature for 30 min.
4. Pellet the beads with a brief centrifugation and transfer approximately half of the supernatant to a new, low-binding tube (label this tube as "IMAC FT").
5. Flatten the end of a 200 µL GELoader pipette tip to retain the IMAC beads. Cut the top of the tip if necessary to tightly fit a syringe.
6. Swirl the IMAC beads in solution to resuspend, then transfer the solution to the GELoader tip. Attach the syringe to the top of the tip and apply positive pressure to expel the liquid into the IMAC FT tube. The IMAC column is now packed.
7. Wash the beads in the GELoader tip with 50 µL of loading buffer A, collect the flow-through in the IMAC FT tube.
8. Wash the IMAC column with 70 µL of elution buffer A (20% ACN in water, 1.5% TFA), collecting the flow-through in the same IMAC FT tube. During this step, the mono-phosphorylated peptides will dissociate from the IMAC beads, retaining only the multiply phosphorylated peptides.
9. Dry the IMAC FT solution in a vacuum centrifuge.
10. Elute the multiply phosphorylated peptides into a new low-binding tube using 80 µL of elution buffer B (1% ammonia in water, pH 11.3).[2]
11. To acidify the eluate for desalting, add 10 µL of 100% formic acid and 2 µL of 10% TFA. The multiply phosphorylated peptides can then be desalted with a C8 stage tip.
12. TiO_2 chromatography can now be used to recover the mono-phosphorylated peptides form the IMAC flow-through. Resuspend the dried IMAC FT contents in 2 µL of 4 M urea and 3 µL of 1% SDS. Add 20 µL of loading buffer B (1 M glycolic acid in 80% ACN, 5% TFA).
13. Prepare a TiO_2 micro-column with TiO_2 beads, based on a C18 microcolumn method [23].

    - To create a micro-column, cut the top of a P200 pipette tip to fit a plastic syringe, which will provide positive pressure.
    - Cut and compress a small disc of 3 M Empore C18 membrane into the bottom of the pipette tip.
    - Transfer the suspended TiO_2 beads to the column, applying pressure with the syringe to pack the column, expelling the liquid.
    - Add 40 µL of 0.1% formic acid to equilibrate the column. Apply just enough pressure to remove excess liquid; do not let the column dry out.

14. Transfer the resuspended peptides to the micro-column.
15. Wash the TiO_2micro-column twice with 50 µL of loading buffer B and once with 30 µL of wash buffer (80% ACN, 5% TFA). Discard the flow-through.

---

[2] This buffer can be more easily made by adding 960 µL of water to 40 µL of a 25% ammonia in water solution.

16. Elute the mono-phosphorylated peptides using 50 μL of ammonium water (5%, pH 11.3) followed by adding 5 μL of 30% ACN to elute possible phosphopeptides bound to the C18 disk. Acidify the eluent with 5 μL of 100% formic acid before desalting.
17. Dry the sample in a vacuum centrifuge and store at −80 °C until mass spectrometry analysis (Fig. 2.3).

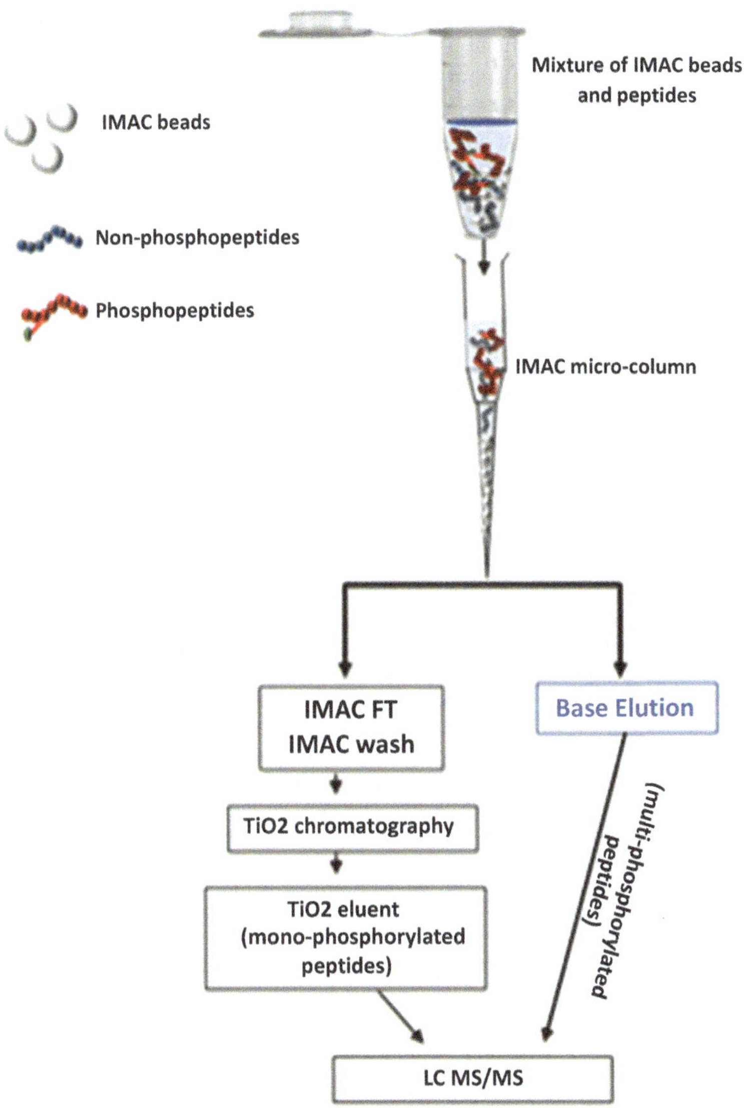

Fig. 2.3 SIMAC phosphoenrichment protocol workflow

2.7 Conclusion

The overall understanding of all the changes provoked in the proteome by a pathology, treatment, or epigenetic modification is also dependent on the capacity to identify post-translational modifications. Many biological processes are controlled by protein phosphorylation, and changes in the phosphorylation state of these proteins can disturb signaling pathways and metabolic processes, even leading to cell death. Phosphopeptide enrichment is crucial when studying the phosphoproteome to overcome issues such as stoichiometric imbalances between the peptide populations and ion suppression caused by the non-phosphorylated peptides, which hides and masks the true quantity of the phosphorylated peptides. Here, several effective methods currently in use to perform phosphopeptide enrichments were presented, allowing the reader to choose the most appropriate method based on the needs of their study.

References

1. Graves, J. D., & Krebs, E. G. (1999). Protein phosphorylation and signal transduction. *Pharmacology & Therapeutics, 82*, 111–121.
2. Thompson, S. M., et al. (2013). Impact of pre-analytical factors on the proteomic analysis of formalin-fixed paraffin-embedded tissue. *PROTEOMICS – Clinical Applications, 7*, 241–251.
3. Hubbard, M. J., & Cohen, P. (1993). On target with a new mechanism for the regulation of protein phosphorylation. *Trends in Biochemical Sciences, 18*, 172–177.
4. Beausoleil, S. A., et al. (2004). Large-scale characterization of HeLa cell nuclear phosphoproteins. *Proceedings of the National Academy of Sciences of the United States of America, 101*, 12130–12135.
5. Beausoleil, S. A., Villén, J., Gerber, S. A., Rush, J., & Gygi, S. P. (2006). A probability-based approach for high-throughput protein phosphorylation analysis and site localization. *Nature Biotechnology, 24*, 1285–1292.
6. Ficarro, S. B., et al. (2009). Magnetic bead processor for rapid evaluation and optimization of parameters for phosphopeptide enrichment. *Analytical Chemistry, 81*, 4566–4575.
7. Olsen, J. V., et al. (2006). Global, in vivo, and site-specific phosphorylation dynamics in signaling networks. *Cell, 127*, 635–648.
8. McNulty, D. E., & Annan, R. S. (2008). Hydrophilic interaction chromatography reduces the complexity of the phosphoproteome and improves global phosphopeptide isolation and detection. *Molecular & Cellular Proteomics, 7*, 971–980.
9. Reinders, J., & Sickmann, A. (2005). State-of-the-art in phosphoproteomics. *Proteomics, 5*, 4052–4061.
10. Thingholm, T. E., Larsen, M. R., Ingrell, C. R., Kassem, M., & Jensen, O. N. (2008). TiO2-based Phosphoproteomic analysis of the plasma membrane and the effects of phosphatase inhibitor treatment. *Journal of Proteome Research, 7*, 3304–3313.
11. Ficarro, S. B., et al. (2002). Phosphoproteome analysis by mass spectrometry and its application to Saccharomyces cerevisiae. *Nature Biotechnology, 20*, 301–305.
12. Andersson, L., & Porath, J. (1986). Isolation of phosphoproteins by immobilized metal (Fe3+) affinity chromatography. *Analytical Biochemistry, 154*, 250–254.
13. Yue, X., Schunter, A., & Hummon, A. B. (2015). Comparing multi-step IMAC and multi-step TiO2 methods for Phosphopeptide enrichment. *Analytical Chemistry, 87*, 8837–8844.

14. Larsen, M. R., Thingholm, T. E., Jensen, O. N., Roepstorff, P., & Jørgensen, T. J. D. (2005). Highly selective enrichment of phosphorylated peptides from peptide mixtures using titanium dioxide microcolumns *. *Molecular & Cellular Proteomics, 4*, 873–886.
15. Bodenmiller, B., Mueller, L. N., Mueller, M., Domon, B., & Aebersold, R. (2007). Reproducible isolation of distinct, overlapping segments of the phosphoproteome. *Nature Methods, 4*, 231–237.
16. Neville, D. C. A., et al. (1997). Evidence for phosphorylation of serine 753 in CFTR using a novel metal-ion affinity resin and matrix-assisted laser desorption mass spectrometry. *Protein Science, 6*, 2436–2445.
17. Posewitz, M. C., & Tempst, P. (1999). Immobilized Gallium(III) affinity chromatography of Phosphopeptides. *Analytical Chemistry, 71*, 2883–2892.
18. Palmisano, G., et al. (2010). Selective enrichment of sialic acid-containing glycopeptides using titanium dioxide chromatography with analysis by HILIC and mass spectrometry. *Nature Protocols, 5*, 1974–1982.
19. Sano, A., & Nakamura, H. (2004). Chemo-affinity of titania for the column-switching HPLC analysis of phosphopeptides. *Analytical Sciences International Journal of Japan Society for Analytical Chemistry, 20*, 565–566.
20. Thingholm, T. E., Jørgensen, T. J. D., Jensen, O. N., & Larsen, M. R. (2006). Highly selective enrichment of phosphorylated peptides using titanium dioxide. *Nature Protocols, 1*, 1929–1935.
21. Jensen, S. S., & Larsen, M. R. (2007). Evaluation of the impact of some experimental procedures on different phosphopeptide enrichment techniques. *Rapid Communications in Mass Spectrometry, 21*, 3635–3645.
22. Thingholm, T. E., Palmisano, G., Kjeldsen, F., & Larsen, M. R. (2010). Undesirable charge-enhancement of isobaric tagged Phosphopeptides leads to reduced identification efficiency. *Journal of Proteome Research, 9*, 4045–4052.
23. Gobom, J., Nordhoff, E., Mirgorodskaya, E., Ekman, R., & Roepstorff, P. (1999). Sample purification and preparation technique based on nano-scale reversed-phase columns for the sensitive analysis of complex peptide mixtures by matrix-assisted laser desorption/ionization mass spectrometry. *Journal of Mass Spectrometry, 34*, 105–116.

Chapter 3
Post-Translational Modifications During Brain Development

Bradley J. Smith and Victor Corasolla Carregari

Abstract Several classes of post-translational modifications (PTMs) regulate various processes that occur during neurodevelopment. The first of these processes is the regulation of the cytoskeleton and cytoskeleton-associating proteins, responsible for the stability, reorganization, and binding of microtubules and actin filaments. Dysregulations in these PTMs lead to dysregulated brain volume and composition, structural defects, behavioral defects, and dendrite growth. The second class of processes involves gene regulation, from chromatin modulation to protein turnover and degradation. Proper gene expression during neurodevelopment is critical to ensure correctly matured cells; dysregulation of PTMs in these pathways leads to various altered morphological and behavioral phenotypes. The third class of processes that are affected by PTMs is cell signaling and signal transduction, vital to cell migration and axonal guidance. Neurodevelopment is a complex sequence of spatially and temporally regulated processes, and PTMs play important roles in this regulation. Most of the known modifications have yet to be studied in depth and much remains undiscovered about their roles in neurodevelopment and otherwise.

Keywords Post-translational modifications · Neurodevelopment · Neurodevelopmental disorders · Brain development

3.1 Biological Steps of Neurodevelopment and Post-translational Modifications

Multiple categories of biological processes are affected and controlled by post-translational modifications during neurodevelopment. During the formation and maturation of the brain, many simultaneous processes occur and must be tightly

B. J. Smith · V. C. Carregari (✉)
Laboratory of Neuroproteomics, Department of Biochemistry and Tissue Biology, Institute of Biology, University of Campinas (UNICAMP), Campinas, Brazil

V. Corasolla Carregari (ed.), *Understanding PTMs in Neurodegenerative Diseases*, Advances in Experimental Medicine and Biology 1382,
https://doi.org/10.1007/978-3-031-05460-0_3

regulated to ensure that no dysregulations in morphology or function arise. Neurons become more specialized, migrate to proper brain regions, and grow axons and dendrites; glial cells quickly form a growth matrix and protect and provide energy support to neurons; and synaptic pruning and myelination occur to trim down excess neurons and protect neurons [1].

The biological functions that will be focused on in this review involve the cytoskeleton, crucial for cell migration, axon guidance, and myelination; gene regulation, including chromatin modulation, transcription factors, mRNA binding proteins, and protein turnover; and signaling pathways, which are vital for cell and axon migration. Each of these functions has been proven to be modulated by one or more post-translational modifications to regulate protein-protein interactions, localization, gene expression, and signal transduction, reinforcing the critical role of PTMs in neurodevelopment.

Previous review articles have focused on specific facets of post-translational modification in the brain such as the roles of lysine acetyltransferases and deacetylases in neurodevelopment [2], histone acetylation in neurodevelopment [3], glycosylation in axonal guidance [4], and PTMs in various types of neuropsychiatric diseases during and after neurodevelopment [5]. In this review, associations between PTMs and the cytoskeleton, gene transcription and protein translation, and signaling pathways will be detailed with a focus on neurodevelopment to provide a concise reference point for future studies involving PTMs and a springboard for investigating the diseases that result from dysregulations in these regulatory pathways, as well as reiterating the potential of this poorly studied field.

3.2 Post-translational Modifications During Neurodevelopment

3.2.1 Cytoskeletal Proteins

One of first links between neurodevelopment and post-translational modifications was suggested to exist in 1990 when multiple charge variants of class III β-tubulin were seen in neurons, but not in tissue from the testis, another known cell type that expresses class III β-tubulin [6]. At that time, the identity and exact sites of the modifications were still unknown. Over the next two decades, microtubules (both α- and β-tubulin) were found to be modified by acetylation, (de)tyrosination, Δ2 modification, polyglutamylation, polyglycylation, palmitoylation, and phosphorylation, affecting their localization and biological function [7]. Post-translationally modifying tubulin subunits is now known to generate subnetworks of tubulin that possess specialized functionality, and this affects microtubule formation dynamics and biological pathways such as neuronal differentiation, plasticity, intracellular transport, and brain function [8] (Fig. 3.1a).

One modification of tubulin in particular was highlighted when mice with suppressed tubulin-tyrosine-ligase, a critical enzyme in the detyrosination/tyrosination cycle, were found to exhibit anomalies in brain morphology and did not survive

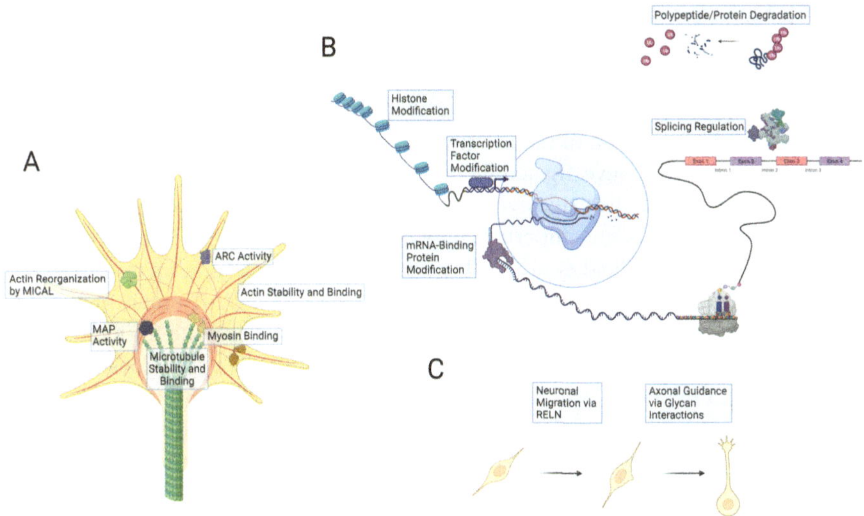

Fig. 3.1 (**a**) Post-translationally modifying tubulin (**b**) Neurodevelopmental processes regulation (**c**) Axonal guidance for neuron migration during neurodevelopment

after birth [9]. Moreover, another detyrosination pathway protein (small vasohibin-binding protein, SVBP) has also been implicated in microtubule structure and neuronal physiology, both in a mouse model and in humans with natural SVBP biallelic inactivation [10]. In that study, SVBP-knockout mice showed a decrease in brain volume, structural defects, white matter loss, and mild behavioral defects, while families with SVBP inactivation showed brain abnormalities, microcephaly, intellectual disability, and delays in gross motor and speech development. Finally, hypothyroidism [11], phenylketonuria [12], and neurodegeneration [13–16] have all been hypothesized to be associated with dysfunctions in tubulin (de)tyrosination. Altogether, these data provide strong evidence that detyrosination/tyrosination regulation plays an important role in neuronal plasticity and brain development and connectivity (reviewed by Moutin et al. 2020).

Acetylation and glutamylation of tubulin have also been studied in relation to brain function and dysfunction. Mice that lack in sensory neurons the protein α-TAT1, the major acetyltransferase for α-tubulin in mice [17], show a great reduction in mechanical stimulation detection and do not show sensitivity to mechanical touch or pain due to decreases in cellular elasticity [18]. These mice also exhibit alterations in the organization of the dentate gyrus, septum, and striatum as well as enlarged lateral ventricles in the forebrain [19, 20]. As such, tubulin acetylation may have an important role in mechanosensitivity and brain development [8]. As for polyglutamylation, this PTM pathway has been discovered to be directly linked to neuronal survival in the central nervous system, peripheral nerves, and spinal motor neurons [21–24]. In humans, biallelic inactivation of CCP1 (a deglutamylase, cytosolic carboxypeptidase 1) directly leads to infantile-onset neurodegeneration [23,

25, 26]. Moreover, glutamylation has been found to be the principal regulating factor of spastin, a microtubule-cleaving enzyme [27, 28], thus being tentatively implicated in hereditary spastic paraplegia [29] and Alzheimer's disease and other tauopathies [30]. Neuron survival is therefore dependent on, at least to a certain degree, polyglutamylation of tubulin (reviewed by Moutin et al. 2020).

Fewer studies have investigated the role of actin modifications in neurodevelopment; however, due to the extensive network of actin filaments in neurons, actin and its interacting partners are important targets for studying PTMs in neurodevelopment (reviewed by Varland et al. 2019) [31]. More specifically, actin turnover rates are regulated in part by phosphorylation, controlling dendritic spine maturation [32]. While not exclusive to neurodevelopment, actin also has a site for methylation whose removal is required for myosin II binding and effective cytokinesis and cell division [33]. Finally, nuclear actin, hypothesized to facilitate chromatin remodeling and therefore gene transcription, can be SUMOylated at two different sites and is thought to help regulate nuclear trafficking and nuclear actin structure [34, 35].

Beyond the core cytoskeletal proteins, microtubule associated proteins MAP1B and MAP2 have long been known to be differentially phosphorylated during neurodevelopment [36, 37]. Over 40 years later, at least 18 sites of MAP2 phosphorylation have been documented that alter its function, some of which have been implicated in schizophrenia [38], a neurodevelopmental and synaptogenetic disorder [39, 40]. The phosphorylation of MICAL1 has been shown to increase its activity and promote efficient and rapid reorganization of actin [41], a process that is important during dendrite outgrowth and neuronal development. Lastly, SUMOylation, ubiquitination, and phosphorylation of ARC (activity-regulated cytoskeletal association protein), have been found to have links with synaptic plasticity [42].

Overall, there is significant evidence that post-translational modifications affect microtubule dynamics and allow for the modulation of multiple neurodevelopmental processes. Moreover, dysregulations in these modification pathways have the potential, and in some cases have been proven, to lead to pathologies. Understanding how different PTMs modify the activity, stability, binding, and localization of cytoskeletal proteins is crucial to understanding and treated related neurodevelopmental dysregulations and disorders.

3.2.2 Gene Transcription and Protein Translation

Several carefully regulated processes are behind the process of a sequence of DNA turning into a fully functional protein, and each step of this journey has the potential to be regulated by post-translational modifications. Neurodevelopment is a finely tuned process that requires both spatial and temporal regulation to properly express genes and allow cells to mature. Several instances of dysregulations in these processes are known to lead to neurodevelopmental disorders and dysregulated phenotypes (Fig. 3.1b).

Histone modifications are one of the earliest regulatory mechanisms, controlling chromatin density and access to gene sequences. Histone acetylation and methylation are modified by prenatal exposure to alcohol, as seen in human and macaque cells [43]. These changes are hypothesized to affect gene transcription to the point of inducing an immature profile of neural cells that can lead to the morphological and behavioral deficits and even fetal death that are observed in fetal alcohol spectrum disorder [44, 45], one of the most common causes of abnormalities in neurodevelopment [46]. In other studies, histone deacetylase activity has been shown to be crucial for neuronal differentiation [47] and the importance of histone acetylation regulation continues throughout neurodevelopment into adulthood [48]. Individual targets of histone acetylation and methylation control the repression of pluripotent, neural, astrocytic, and non-neuronal genes to ensure proper cell maturation [49, 50]. Lastly, certain stimuli during critical periods throughout post-natal neurodevelopment have been suggested to have lasting consequences via the modulation of histone deacetylase (HDAC5) activity [51].

Transcription factors are also vital for proper gene expression and can me modulated by PTMs to affect their binding efficiency and targets. In non-neuronal cells for example, the transcription factor NRSF (neuron-restrictive silencing factor) is responsible for repressing neuronal differentiation and activation and neurogenesis [52] by binding to specific gene promoters and associating with repressor complexes [48, 53], which includes a histone deacetylase complex (coREST/HDAC2) [54]. OLIG1 is another transcription factor that plays a critical role in oligodendrocyte development and whose nuclear export sequence contains a acetylation site that regulates its nuclear-to-cytoplasmic translocation [55]. Processes that occur later in neurodevelopment have also been associated with PTM modulation such as fear-related, long-term memory formation via phosphorylation of the transcription factor CREB [56] and multiple facets of cognition are thought to be affected by chromatin regulation [57].

Once an mRNA sequence has been transcribed, it can bind to various proteins to promote or repress its translation. In one such example, FMRP (Fragile X Mental Retardation Protein) binds to mRNA sequences as a repressor of translation and is critical to synapse function. It can be modified by several PTMs whose interplay, both spatially and temporally, is important for proper neurological functioning; in fact, a loss of function of this protein as well as a missense genetic mutation can both lead to Fragile X syndrome, encompassing intellectual disability, impairments in memory and learning, autistic-like features, and seizures [58].

Finally, polypeptides can be modified through several processes, including splicing, degradation and turnover, and PTMs of their own. SRPK (Ser/Arg-rich splicing factor protein kinase) is a protein that is responsible for splicing regulation and it controls RNF12/RLIM (RING finger ubiquitin ligase E3) by phosphorylation, ensuring proper regulation of neurodevelopmental gene expression by ubiquitin signaling [59]. E3 ubiquitin ligase (UBE3A) has several biological functions in GABAergic neurons, regulating a wide array of pathways, making it critical for neurodevelopment [60]. Various modifications occur at every step of gene expression; while few modifications have been studied in regard to neurodevelopment, many PTMs are known to exist and each plays unique roles in translating endogenous and exogenous stimuli into modulations in gene transcription, reinforcing the need to study more PTMs during critical developmental processes.

3.2.3 Signaling Pathways

Cell migration within the brain and various processes of outgrowth and myelination require signaling pathways to ensure that cells reach their intended targets quickly and efficiently. One pathway for neuronal migration that occurs in multiple brain regions involves extracellular signaling via reelin (RELN), which affects cells by interaction with the membrane protein disabled-1 (DAB1), affecting its expression and phosphorylation [61]. After this discovery, Howell et al. then created a mutated version of DAB1 in mice without tyrosine phosphorylation sites and found that DAB1 phosphorylation, and not protein expression, is critical for correct neuronal migration [62].

Once cells have reached their intended destination, neurons require axonal guidance to ensure that axons make proper connections during neurodevelopment (Fig. 3.1c). A large range of proteins and protein classes are responsible for this critical guidance [63], providing both attractive and repulsive signaling cues. The wide range of glycan structures found the mammalian brain [64], along with the known role of glycosylation in cell adhesion, trafficking, and signaling [65], make glycosaminoglycans and N-linked and O-linked glycans perfect PTMs for cell signaling, and this is seen *in vivo* during axonal guidance (reviewed by Mutalik and Gupton 2021).

3.3 Concluding Remarks

The vast majority of data published on post-translational modifications in neurodevelopment has stuck to the more classical and canonical modifications of acetylation, methylation, and phosphorylation, with few studies venturing out into the effects of other modifications such as glycosylation or ubiquitination. In reality, hundreds of PTMs are known to exist [66]; but studying each of them is resource-intensive, requiring vast amounts of time and funding to study, leading to the need for *in silico* methods to assist in site predictions [67]. As a result, many investigations have been carried in a more targeted fashion, seeking information about a particular modification or a specific, modified protein. This is further complicated by the difficulty of manipulating or otherwise obtaining cells of the human central nervous system, particularly during stages of neurodevelopment; however, the ability to generate cerebral organoids from human pluripotent stem cells has revolutionized the field [68].

With the great deal of potential held in PTMs and the regulation of precise and quickly changing processes such as neurodevelopment, the field of proteomics, or more specifically PTMomics, has much to offer. Examples that have been studied without a focus on neurodevelopment include lipidation for protein localization to membranes [69], which can be especially important in neurons with multiple membrane-bound and transmembrane proteins; other types of acylation such as succinylation, malonylation, propionylation, and butyrylation [70], which have important precursors in energy metabolism; pyroglutamylation, which has been studied in relation to tauopathies [71, 72]; and tyrosine sulfation, which occurs in many membrane and secreted proteins [73]. Studying these modifications and others in relation to neurodevelopment can provide insight into the processes that they regulate and how to prevent, diagnose, and treat various neurodevelopmental diseases and disorders.

References

1. Budday, S., Steinmann, P., & Kuhl, E. (2015). Physical biology of human brain development. *Frontiers in Cellular Neuroscience, 9*, 257.
2. Tapias, A., & Wang, Z.-Q. (2017). Lysine acetylation and Deacetylation in brain development and neuropathies. *Genomics, Proteomics & Bioinformatics, 15*, 19–36.
3. Contestabile, A., & Sintoni, S. (2013). Histone acetylation in neurodevelopment. *Current Pharmaceutical Design, 19*, 5043–5050.
4. Mutalik, S. P., & Gupton, S. L. (2021). Glycosylation in axonal guidance. *International Journal of Molecular Sciences, 22*, 5143.
5. Grubisha, M. J., Sweet, R. A., & MacDonald, M. L. (2021). Investigating post-translational modifications in neuropsychiatric disease: The next frontier in human post-mortem brain research. *Frontiers in Molecular Neuroscience, 14*, 145.
6. Lee, M. K., Rebhun, L. I., & Frankfurter, A. (1990). Posttranslational modification of class III beta-tubulin. *PNAS, 87*, 7195–7199.
7. Fukushima, N., Furuta, D., Hidaka, Y., Moriyama, R., & Tsujiuchi, T. (2009). Post-translational modifications of tubulin in the nervous system. *Journal of Neurochemistry, 109*, 683–693.
8. Moutin, M.-J., Bosc, C., Peris, L., & Andrieux, A. (2020). Tubulin post-translational modifications control neuronal development and functions. *Developmental Neurobiology, 81*, 253–272.
9. Erck, C., et al. (2005). A vital role of tubulin-tyrosine-ligase for neuronal organization. *Proceedings of the National Academy of Sciences of the United States of America, 102*, 7853–7858.
10. Pagnamenta, A. T., et al. (2019). Defective tubulin detyrosination causes structural brain abnormalities with cognitive deficiency in humans and mice. *Human Molecular Genetics, 28*, 3391–3405.
11. Poddar, R., & Sarkar, P. K. (1993). Delayed detyrosination of alpha-tubulin from parallel fibre axons and its correlation with impaired synaptogenesis in hypothyroid rat cerebellum. *Brain Research, 614*, 233–240.
12. Ditamo, Y., Dentesano, Y. M., Purro, S. A., Arce, C. A., & Bisig, C. G. (2016). Post-translational incorporation of L-phenylalanine into the C-terminus of α-tubulin as a possible cause of neuronal dysfunction. *Scientific Reports, 6*, 38140.
13. Chen, H., et al. (2015). Meta-analyses on prevalence of selected Parkinson's nonmotor symptoms before and after diagnosis. *Transl Neurodegener, 4*, 1.

14. Gonatas, N. K., Stieber, A., & Gonatas, J. O. (2006). Fragmentation of the Golgi apparatus in neurodegenerative diseases and cell death. *Journal of the Neurological Sciences, 246*, 21–30.
15. Jackson, M. Z., Gruner, K. A., Qin, C., & Tourtellotte, W. G. (2014). A neuron autonomous role for the familial dysautonomia gene ELP1 in sympathetic and sensory target tissue innervation. *Development, 141*, 2452–2461.
16. Pianu, B., Lefort, R., Thuiliere, L., Tabourier, E., & Bartolini, F. (2014). The $A\beta_{1-42}$ peptide regulates microtubule stability independently of tau. *Journal of Cell Science, 127*, 1117–1127.
17. Kalebic, N., et al. (2013). αTAT1 is the major α-tubulin acetyltransferase in mice. *Nature Communications, 4*, 1962.
18. Morley, S. J., et al. (2016). Acetylated tubulin is essential for touch sensation in mice. *eLife, 5*, e20813.
19. Kim, G.-W., Li, L., Ghorbani, M., You, L., & Yang, X.-J. (2013). Mice lacking α-tubulin acetyltransferase 1 are viable but display α-tubulin acetylation deficiency and dentate gyrus distortion. *The Journal of Biological Chemistry, 288*, 20334–20350.
20. Li, L., et al. (2019). ATAT1 regulates forebrain development and stress-induced tubulin hyperacetylation. *Cellular and Molecular Life Sciences, 76*, 3621–3640.
21. Muñoz-Castañeda, R., et al. (2018). Cytoskeleton stability is essential for the integrity of the cerebellum and its motor- and affective-related behaviors. *Scientific Reports, 8*, 3072.
22. Rogowski, K., et al. (2010). A family of protein-deglutamylating enzymes associated with neurodegeneration. *Cell, 143*, 564–578.
23. Shashi, V., et al. (2018). Loss of tubulin deglutamylase CCP1 causes infantile-onset neurodegeneration. *The EMBO Journal, 37*, e100540.
24. Zhou, L., et al. (2018). Deletion of exons encoding carboxypeptidase domain of Nna1 results in Purkinje cell degeneration (pcd) phenotype. *Journal of Neurochemistry, 147*, 557–572.
25. Karakaya, M., et al. (2019). Biallelic variant in AGTPBP1 causes infantile lower motor neuron degeneration and cerebellar atrophy. *American Journal of Medical Genetics. Part A, 179*, 1580–1584.
26. Sheffer, R., et al. (2019). Biallelic variants in AGTPBP1, involved in tubulin deglutamylation, are associated with cerebellar degeneration and motor neuropathy. *European Journal of Human Genetics, 27*, 1419–1426.
27. Lacroix, B., et al. (2010). Tubulin polyglutamylation stimulates spastin-mediated microtubule severing. *The Journal of Cell Biology, 189*, 945–954.
28. Valenstein, M. L., & Roll-Mecak, A. (2016). Graded control of microtubule severing by tubulin Glutamylation. *Cell, 164*, 911–921.
29. Hazan, J., et al. (1999). Spastin, a new AAA protein, is altered in the most frequent form of autosomal dominant spastic paraplegia. *Nature Genetics, 23*, 296–303.
30. Zempel, H., et al. (2013). Amyloid-β oligomers induce synaptic damage via tau-dependent microtubule severing by TTLL6 and spastin. *The EMBO Journal, 32*, 2920–2937.
31. Varland, S., Vandekerckhove, J., & Drazic, A. (2019). Actin post-translational modifications: The Cinderella of cytoskeletal control. *Trends in Biochemical Sciences, 44*, 502–516.
32. Bertling, E., et al. (2016). Actin Tyrosine-53-phosphorylation in neuronal maturation and synaptic plasticity. *The Journal of Neuroscience, 36*, 5299–5313.
33. Li, M.-M., et al. (2013). ALKBH4-dependent demethylation of actin regulates actomyosin dynamics. *Nature Communications, 4*, 1832.
34. Hofmann, W. A., et al. (2009). SUMOylation of nuclear actin. *Journal of Cell Biology, 186*, 193–200.
35. Uda, M., et al. (2015). Sumoylated α-skeletal muscle actin in the skeletal muscle of adult rats. *Molecular and Cellular Biochemistry, 409*, 59–66.
36. Fischer, I., & Romano-Clarke, G. (1990). Changes in microtubule-associated protein MAP1B phosphorylation during rat brain development. *Journal of Neurochemistry, 55*, 328–333.
37. Riederer, B. M., Draberova, E., Viklicky, V., & Draber, P. (1995). Changes of MAP2 phosphorylation during brain development. *The Journal of Histochemistry and Cytochemistry, 43*, 1269–1284.

38. Grubisha, M. J., et al. (2021). MAP2 is differentially phosphorylated in schizophrenia, altering its function. *Molecular Psychiatry*. https://doi.org/10.1038/s41380-021-01034-z
39. Grant, S. G. N. (2012). Synaptopathies: Diseases of the synaptome. *Current Opinion in Neurobiology, 22*, 522–529.
40. Washbourne, P. (2015). Synapse assembly and neurodevelopmental disorders. *Neuropsychopharmacology, 40*, 4–15.
41. Yoon, J., Kim, S. B., Ahmed, G., Shay, J. W., & Terman, J. R. (2017). Amplification of F-actin disassembly and cellular repulsion by growth factor signaling. *Developmental Cell, 42*, 117–129.e8.
42. Carmichael, R. E., & Henley, J. M. (2018). Transcriptional and post-translational regulation of arc in synaptic plasticity. *Seminars in Cell & Developmental Biology, 77*, 3–9.
43. Jarmasz, J. S., et al. (2019). Global DNA methylation and histone posttranslational modifications in human and nonhuman primate brain in association with prenatal alcohol exposure. *Alcoholism: Clinical and Experimental Research, 43*, 1145–1162.
44. Carson, G., et al. (2010). Alcohol use and pregnancy consensus clinical guidelines. *Journal of Obstetrics and Gynaecology Canada, 32*, 1–31.
45. Reynolds, J. N., et al. (2011). Fetal alcohol spectrum disorders: Gene-environment interactions, predictive biomarkers, and the relationship between structural alterations in the brain and functional outcomes. *Seminars in Pediatric Neurology, 18*, 49–55.
46. Popova, S., Lange, S., Burd, L., & Rehm, J. (2012). Health care burden and cost associated with fetal alcohol syndrome: Based on official Canadian data. *PLoS One, 7*, e43024.
47. Marin-Husstege, M., Muggironi, M., Liu, A., & Casaccia-Bonnefil, P. (2002). Histone deacetylase activity is necessary for oligodendrocyte lineage progression. *The Journal of Neuroscience, 22*, 10333–10345.
48. Hsieh, J., & Gage, F. H. (2005). Chromatin remodeling in neural development and plasticity. *Current Opinion in Cell Biology, 17*, 664–671.
49. Hirabayashi, Y., & Gotoh, Y. (2010). Epigenetic control of neural precursor cell fate during development. *Nature Reviews. Neuroscience, 11*, 377–388.
50. Sheikh, B. N. (2014). Crafting the brain – Role of histone acetyltransferases in neural development and disease. *Cell and Tissue Research, 356*, 553–573.
51. Renthal, W., et al. (2007). Histone deacetylase 5 epigenetically controls behavioral adaptations to chronic emotional stimuli. *Neuron, 56*, 517–529.
52. Kuwabara, T., Hsieh, J., Nakashima, K., Taira, K., & Gage, F. H. (2004). A small modulatory dsRNA specifies the fate of adult neural stem cells. *Cell, 116*, 779–793.
53. Ballas, N., & Mandel, G. (2005). The many faces of REST oversee epigenetic programming of neuronal genes. *Current Opinion in Neurobiology, 15*, 500–506.
54. Ballas, N., et al. (2001). Regulation of neuronal traits by a novel transcriptional complex. *Neuron, 31*, 353–365.
55. Dai, J., Bercury, K. K., Jin, W., & Macklin, W. B. (2015). Olig1 acetylation and nuclear export mediate oligodendrocyte development. *The Journal of Neuroscience, 35*, 15875–15893.
56. Lisman, J., Cooper, K., Sehgal, M., & Silva, A. J. (2018). Memory formation depends on both synapse-specific modifications of synaptic strength and cell-specific increases in excitability. *Nature Neuroscience, 21*, 309–314.
57. Day, J. J., & Sweatt, J. D. (2011). Epigenetic mechanisms in cognition. *Neuron, 70*, 813–829.
58. Prieto, M., Folci, A., & Martin, S. (2020). Post-translational modifications of the fragile X mental retardation protein in neuronal function and dysfunction. *Molecular Psychiatry, 25*, 1688–1703.
59. Bustos, F., et al. (2020). Functional diversification of SRSF protein kinase to control ubiquitin-dependent neurodevelopmental signaling. *Developmental Cell, 55*, 629–647.e7.
60. Lopez, S. J., Segal, D. J., & LaSalle, J. M. (2019). UBE3A: An E3 ubiquitin ligase with genome-wide impact in neurodevelopmental disease. *Frontiers in Molecular Neuroscience, 11*, 476.

61. Howell, B. W., Herrick, T. M., & Cooper, J. A. (1999). Reelin-induced tryosine phosphorylation of disabled 1 during neuronal positioning. *Genes & Development, 13*, 643–648.
62. Howell, B. W., Herrick, T. M., Hildebrand, J. D., Zhang, Y., & Cooper, J. A. (2000). Dab1 tyrosine phosphorylation sites relay positional signals during mouse brain development. *Current Biology, 10*, 877–885.
63. McCormick, L. E., & Gupton, S. L. (2020). Mechanistic advances in axon pathfinding. *Current Opinion in Cell Biology, 63*, 11–19.
64. Lee, J., et al. (2020). Spatial and temporal diversity of glycome expression in mammalian brain. *PNAS, 117*, 28743–28753.
65. Dong, X., et al. (2018). Advances in mass spectrometry-based glycomics. *ELECTROPHORESIS, 39*, 3063–3081.
66. UniProt Consortium. *Controlled vocabulary of posttranslational modifications (PTM)*. (2021).
67. Ramazi, S., & Zahiri, J. (2021). Post-translational modifications in proteins: Resources, tools and prediction methods. *Database, 2021*.
68. Lancaster, M. A., & Knoblich, J. A. (2014). Generation of cerebral organoids from human pluripotent stem cells. *Nature Protocols, 9*, 2329–2340.
69. Ray, A., Jatana, N., & Thukral, L. (2017). Lipidated proteins: Spotlight on protein-membrane binding interfaces. *Progress in Biophysics and Molecular Biology, 128*, 74–84.
70. Xu, J.-Y., et al. (2018). Protein acylation is a general regulatory mechanism in biosynthetic pathway of acyl-CoA-derived natural products. *Cell Chemical Biology, 25*, 984–995.e6.
71. Mandler, M., et al. (2014). Pyroglutamylated amyloid-β is associated with hyperphosphorylated tau and severity of Alzheimer's disease. *Acta Neuropathologica, 128*, 67–79.
72. Moro, M. L., et al. (2018). Pyroglutamate and Isoaspartate modified amyloid-Beta in ageing and Alzheimer's disease. *Acta Neuropathologica Communications, 6*, 3.
73. Stone, M. J., Chuang, S., Hou, X., Shoham, M., & Zhu, J. Z. (2009). Tyrosine sulfation: An increasingly recognised post-translational modification of secreted proteins. *New Biotechnology, 25*, 299–317.

Chapter 4
Aberrant Protein Glycosylation in Brain Cancers, with Emphasis on Glioblastoma

Livia Rosa-Fernandes, Sueli Mieko Oba-Shinjo, Janaina Macedo-da-Silva, Suely Kazue Nagahashi Marie, and Giuseppe Palmisano

Abstract Aberrant glycosylation has been associated with several processes of tumorigenesis from cell signaling, migration and invasion, to immune regulation and metastasis formation. The biosynthesis of glycoconjugates is regulated through concerted and finely tuned enzymatic reactions. This includes the levels and activity of glycosyltransferases and glycosidases, nucleotide sugar metabolism, substrate availability, epigenetic condition, and cellular functional state. Glioblastoma (GBM) is the most aggressive brain tumor, frequently occurring in adults with overall survival not surpassing 17 months after diagnosis. GBM has been classified by the World Health Organization (WHO) as a grade 4 astrocytoma and stratified into G-CIMP, proneural, classical, and mesenchymal subtypes. Several biomolecular features associated with GBM aggressiveness have been elucidated; however, more studies are needed to elucidate the role of glycosylation in GBM pathology, looking at their potential as cancer targets. Here, we focus on the alteration of genes involved in protein N- and O-linked glycosylation in GBM. Specifically, the mRNA levels of glycogenes were analyzed using astrocytoma-TCGA-RNAseq datasets from public repositories. A total of 68 genes were differentially regulated in the most aggressive, mesenchymal subtype of GBM compared to the proneural and classical subtypes, and the expression of these genes was compared to normal brain tissues. Among them, we focused on 38 genes coding for proteins that belong to: 1) asparagine

Livia Rosa-Fernandes, Sueli Mieko Oba-Shinjo and Janaina Macedo-da-Silva contributed equally with all other contributors.

L. Rosa-Fernandes · J. Macedo-da-Silva · G. Palmisano (✉)
Department of Parasitology, Institute of Biomedical Sciences, University of São Paulo, São Paulo, SP, Brazil
e-mail: palmisano.gp@usp.br

S. M. Oba-Shinjo · S. K. N. Marie (✉)
Laboratory of Cellular and Molecular Biology (LIM 15), Department of Neurology, Faculdade de Medicina, University of São Paulo, São Paulo, SP, Brazil
e-mail: sknmarie@usp.br

V. Corasolla Carregari (ed.), *Understanding PTMs in Neurodegenerative Diseases*, Advances in Experimental Medicine and Biology 1382,
https://doi.org/10.1007/978-3-031-05460-0_4

glycosylation (ALG); 2) glycosyltransferases (B3T, B4T); 3) fucosyltransferase (FUT); 4) acetylgalactosaminyltransferases (GALNT); 5) hexosaminidase (HEX); 6) mannosidase (MAN); 7) acetylglucosaminyltransferase (MGAT); 8) sialidase or neuraminidase (NEU); 9) solute carrier 35 family (SLC); and 10) sialyltransferase (ST). The differential expression of some genes was already reported in several solid tumors; however, several of them were found to be dysregulated in GBM for the first time. These data represent an important starting point to perform further orthogonal and functional validations to pinpoint the role of these glycogenes in GBM as diagnostic and therapeutic targets.

Keywords Protein glycosylation · Glycogenes · Brain cancers · Glioblastoma · Epithelial-mesenchymal transition · RNAseq

4.1 Introduction

The glycoconjugate repertoire of a cell is formed by the covalent linkage between carbohydrates, or glycans, with lipids, proteins, and RNA. These biomolecules are present on the cell surface and in extracellular and intracellular spaces [1]. To date, more than 180 glycosyltransferase genes have been discovered to be involved in the biosynthesis of glycoconjugates [2, 3]. Proteins can be glycosylated by the covalent attachment of an (oligo)saccharide chain to a polypeptide backbone. Two main linkages are present in mammalian cells: an N-linkage formed between the terminal GlcNAc of the glycan core and the amidic nitrogen of an asparagine residue (Asn) and an O-linkage formed between the first GalNAc residue to the hydroxyl group of a serine (Ser) or threonine (Thr) residue [4].

N-glycan attachment to a glycoprotein occurs at the consensus peptide sequence Asn-X-Ser/Thr, with X being any amino acid except proline, and serves to influence protein solubility, stabilize protein structure and mediate receptor ligand interactions [5, 6]. Protein glycosylation occurs in the endoplasmic reticulum (ER) and Golgi compartment, where the concerted and finely tuned actions of glycosyltransferases and glycosidases occur, shaping the cellular glycome. The specific glycoconjugate repertoire is modulated by the status of nucleotide sugar metabolism, substrate availability, epigenetic condition, and cellular functional state. Protein glycosylation is critical for several biological processes and, in its aberrant form, can induce and dictate a pathological state. Consequently, an altered expression of the glycosylation biosynthetic machinery can be associated with a disease state.

Cumulative evidence has shown the involvement of protein glycosylation in basic molecular and cellular biological processes during the development and progression of cancer. Aberrant glycosylation has been associated with tumor cell signaling, migration and invasion, cell-cell and cell-extracellular matrix interactions, angiogenesis, immune regulation, and metastasis formation [4].

The present review is focused on the impact of protein glycosylation in brain tumors, particularly in glioblastoma (GBM). Tumors of the central nervous system are less frequent than other locations, representing about 3% of all tumors; but they are among the most aggressive [7, 8]. Gliomas, which originate from glial cells or their precursors, represent more than 80% of brain tumors [9, 10]. GBM is the most aggressive brain malignancy and frequently occurs in adults [11]. Despite continuous efforts to search for new therapeutic strategies [12], the outcome of GBM patients remains dismal, not exceeding 17 months of survival after diagnosis [13]. GBM is classified by World Health Organization (WHO) as a grade IV astrocytoma, and has been stratified into subtypes according to their molecular profiles, being G-CIMP, proneural, classical, and mesenchymal [14, 15], which predict partially the clinical outcome. The proneural subtype, characterized by the presence of *IDH* mutation, has been associated with a better prognosis [16], while the mesenchymal subtype, with *NF1* or *RB1* mutations, tends to present the worst outcome [17].

The aggressiveness of GBM is attributed to several biological characteristics of the tumor cells, including the hyperactivation of proliferation signaling pathways through receptor tyrosine kinases, loss of cell adhesion leading to tumor cell migration and invasion of surrounding normal tissue, maintenance of tumor stem cells, activated epithelial mesenchymal transition (EMT), and evasion of the immune response. Modifications of the glycosylation pattern of tumor cells and other cells or proteins of the tumor microenvironment impact each of these characteristics of malignancy, which have been analyzed and explored for therapeutic purposes. In the next sections, we will highlight the influence of glycosylation on important aspects of malignancy.

Cell Membrane Receptors Malignant transformation of GBM is associated with aberrant expression of glycans on the cell surface, as N-glycosylation may affect receptor structure, stability, and function. About 90% of GBMs have genomic mutations, amplifications, or deletions that activate pathways mediated by tyrosine kinase receptors [18]. The epidermal growth factor receptor (EGFR) is one of the receptor tyrosine kinases that are upregulated in several tumors, including GBM, mainly the classical subtype. The structural modification of EGFR by N-glycosylation may promote dimerization [19, 20] and alter cross-cell signal transduction [21], binding affinity of growth factors and of monoclonal antibodies [21], conformation in the cell membrane [22], downstream signaling to cell growth, and sensitivity to gefitinib [23].

Cell Adhesion/Migration/Invasion Tumor invasiveness is the characteristic of GBM that prevents total tumor resectability and determines its inevitable recurrence. Tumor cell motility depends on the dynamic cell-extracellular matrix (ECM) and cell-cell interactions. E-cadherin participates in cell adhesion, and N-glycosylation changes its molecular structure, thereby destabilizing adherens junctions [24]. Moreover, the degree of N-glycan branching of E-cadherin is inversely related to the establishment of cell adhesion [24]. E-cadherin core fucosylation activates the FAK/integrin pathway, increasing metallopeptidases 2 and 9 (*MMP-2*, *MMP-9*), which enhance the adhesion, migration, and invasion of cancer cells [25]. Additionally, E-cadherin O-glycosylation in the cytoplasm blocks its transport during the process of apoptosis, leading to ER stress and decreases intercellular adhesion [26]. Inversely, N-glycosylation of N-cadherin at Asn402 controls stability and plays a role in cell-cell adhesion of glioma cells [27].

Integrins are the major components of the cell-ECM interaction and they contain more than 20 potential glycosylation sites [28]. Their N-glycan structures are essential for integrin heterodimerization, conformational stability, localization on the cell membrane, and interaction with ligands [29]. Core fucosylation of α3β1 integrin can induce laminin 5-induced migration and intracellular signal transduction [30].

Increased sialylation of tumors promotes metastatic events, immune evasion, invasiveness, and cancer migration. The terminal position of sialic acid residues turns the cell surface negatively charged inducing repulsion between cell membrane molecules. Thus, malignancy-associated hypersialylation promotes the escape of cells from the tumor mass and invasion into the surrounding tissue. In fact, an *in vitro* model of glioma study demonstrated that polysialic acid cells exhibited increased invasion of the corpus callosum [31]. Therefore, the sialic acid-rich glycocalyx in highly malignant glioma cells may be a negative prediction marker compared to low-grade malignancies [32]. Additionally, mannose trimming by different mannosidases at the initial steps of N-glycosylation leads to increased diversity of N-glycans, enhancing further steps of fucosylation and sialylation. Thus, the expression level of Golgi α-mannosidase affects adhesive properties, being associated with the outcome of breast cancer patients [33].

Cell Stemness GBM-like stem cells (GSCs) are considered one of the main reasons for tumor recurrence, as they are behind the rapid growth, therapeutic resistance, and clinical recurrence of these fatal tumors [34, 35]. Activation of the Wnt/β-catenin signaling pathway upregulates the expression level of *FUT8*, thereby inducing an increase of tumor cell stemness [36]. Overexpression of *TRIM8* in GBMs led to an upregulation of p-STAT3, c-MYC, SOX2, NESTIN, and CD133 and enhanced GSC auto-renewal [35]. CD133, the self-renewing cancer stem cell marker in a variety of solid tumors including GBM, may drive intra-tumoral heterogeneity. CD133 N-glycosylation is important for cell surface recognition of the epitope. Important enzymes of N-glycan biosynthesis, such as *ALG1, ALG5, ALG8, ALG10, GANC, MGAT3, MGAT4C ST3GAL4, ST6GAL1, and ST6GAL2*, are involved in CD133 glycosylation and stability [37].

Epithelial-Mesenchymal Transition EMT is regulated by a series of signaling pathways and is associated with cancer invasion and progression [38]. Glycosylation changes are intimately linked to the EMT process [39, 40], which has been demonstrated in a whole-genome sequencing data analysis of *in vitro* pancreatic cancer models of EMT [41]. Several enzymes involved in glycosaminoglycan sulfation, desulfation, O-linked glycosylation, and sialylation were detected in this study, highlighting the importance of glycogenes as targets of EMT. Similarly, core fucosylation was also reported as highly upregulated during the process of EMT, induced by TGF-β [42], as well as by activation of the Wnt/β-catenin signaling in breast cancer cells [36]. Additionally, mannosyltransferases have been associated with EMT of glioma cancer stem cells [43].

Immunoresponse Inhibition of core fucosylation can reduce the expression of PD-1 on the cell surface, by enhancing its ubiquitination and degradation [44], and can more effectively eradicate the tumors by blocking PD-1/PD-L1-mediated tumor immune escape [45]. Exosomes derived from GBMs have glycan motifs that function as inappropriate tumor antigens that are unable to activate the immune system. Removal of these immunosuppressive glycans and insertion of a high-affinity glycoprotein ligand for antigen presenting cells considerably reduced immunosuppression [46]. Hypersialylation modulates immunological synapses and potentiates tumor immune evasion by engaging Siglecs (sialic acid-binding immunoglobulin superfamily lectins) carrying an immunoreceptor tyrosine-based inhibition motif and eliciting inhibitory intracellular signaling in immune cells [47]. In the brain, the sialoglycans are recognized by Siglecs on the microglia and infiltrating peripheral immune cells. Particularly, Siglec-11, a member of the CD33/Siglec-3-related Siglecs, is expressed in microglia [48] and recognizes its only known ligand, polysialylated neural adhesion molecules, present on malignant cells. Then, glioma immune escape may be facilitated by inhibitory Siglecs interacting with sialic acid on tumor cells and triggering a decreased immune response.

Potential Therapeutic Targets 2-fluoro-L-fucose (2FF), a core fucosylation inhibitor, reduced the fucosylation level of core membrane glycoproteins and inhibited liver cancer progression [49]. In another study, 2FF in combination with anti-PDL1 enhanced the therapeutic efficacy in breast cancer [50]. A novel small-molecule inhibitor (NGI-1) that inhibits the glycosylation and phosphorylation of multiple receptor tyrosine kinases in GBM increased tumor cell sensibility to radiation, enhanced cytotoxic chemotherapy, and significantly reduced tumor growth *in vivo* [51, 18]. The inhibition of core fucosylation of IgG FC enhances binding affinity to FcγRIIIa [5] and increases the antibody-dependent cytotoxicity of natural killer lymphocytes [52].

Additionally, modifications of sialylation of cancer cells can modulate the effectiveness of anti-cancer therapy by blocking the interactions between protein ligands and immune checkpoints, which normally activate immune cells to destroy the tumor cells. Moreover, when nanoparticles containing anti-cancer compounds are coated in a sialic layer, it can facilitate their crossing the blood-brain barrier [53], enhance their delivery to tumor cells, and increase therapeutic efficacy [54]. Nanoformulated sialyltransferase inhibitors, such as fluorinated sialic derivatives, may alter sialylation patterns in tumor cells and affect sialic acid-dependent adhesion, migration, and viability. The blockage of the sialic acid-inhibitory Siglec axis is a promising strategy in the host defense against glioma cells [55].

Genes that code for enzymes that are involved in N- and O-glycosylation of proteins (here named glycogenes) in human cells were analyzed using public databases. Searching for additional candidates for GBM therapy, we analyzed the differentially expressed glycogenes in RNAseq datasets of the Cancer Genome

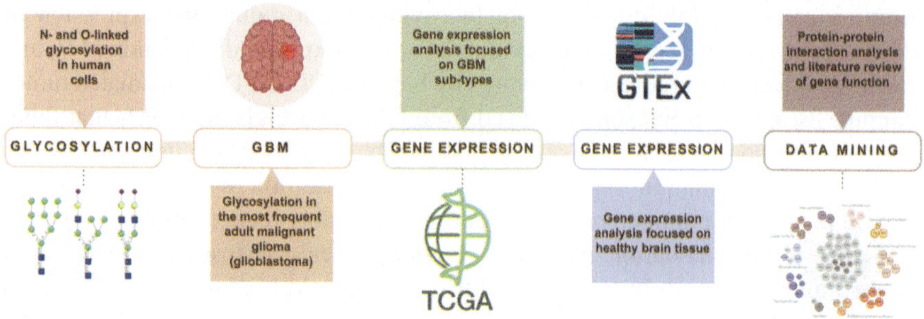

Fig. 4.1 Summary of the analysis workflow. We searched for expressed human glycogenes in RNAseq datasets from public repositories: TCGA (astrocytoma) and GTEx (non-neoplastic brain tissue). The expression data were normalized and compared between the different datasets using stringent statistical analyses. The differentially regulated glycogenes were selected for data mining and reviewed based on their families

Atlas (TCGA) (astrocytoma) and the Genotype-Tissue expression (GTEx) (non-neoplastic brain tissue) (Fig. 4.1). Astrocytoma-TCGA-RNAseq datasets was downloaded (Genomics Data Commons Data Portal - https://portal.gdc.cancer.gov/) and normalized by DESeq R software. Among the 155 glycogenes related to protein glycosylation that were differentially expressed in GBM compared to lower grade astrocytomas (grade 2 and grade 3), 68 genes were differentially expressed in the most aggressive mesenchymal subtype of GBM compared to proneural and classical subtypes (Fig. 4.2). Among them, thirty-eight genes code for members of 10 relevant enzyme families: 1) asparagine glycosylation (ALG); 2) glycosyltransferases (B3T, B4T); 3) fucosyltransferases (FUT); 4) acetylgalactosaminyltransferases (GALNT); 5) hexosaminidases (HEX); 6) mannosidases (MAN); 7) acetylglucosaminyltransferases (MGAT); 8) sialidases or neuraminidases (NEU); 9) the solute carrier 35 family (SLC); and 10) sialyltransferases (ST).

Fig. 4.2 (continued) and normalized by DESeq R software. The RPKM values were normalized by z-score. Sixty-eight glycogenes were differentially expressed among 114 GBM cases, consisting of 28 proneural (PR), 38 classical (CS), and 48 mesenchymal (MS) molecular subtypes. Statistical significances were shown as * $p < 0.05$, ** $p < 0.01$, *** $p < 0.001$ by Kruskal-Wallis (K-W) test, and post-hoc Dunn's test as ¶ $p < 0.05$, ¶¶ $p < 0.01$, ¶¶¶ $p < 0.001$ for PN vs MS comparison, and ǂ $p < 0.05$, ǂǂ $p < 0.01$, ǂǂǂ $p < 0.001$ for CS vs MS comparison. The gene expression levels of MS GBM were compared to 207 non-neoplastic brain tissue sets from the GTEx database by the Gene Expression Profiling Interactive Analysis 2 (GEPIA 2) online tool (http://gepia2.cancer-pku.cn), and the genes differentially expressed were marked with @. Up-regulated gene expression levels are in red and down-regulated in blue. The members of the families are identified with specific colors as shown for ALG: asparagine N-linked glycosylation, B3T, B4T: glycosyltransferase, FUT: fucosyltransferase, GALNT: acetylgalactosaminyltransferase, HEX: hexosaminidase, MAN: mannosidase, MGAT: acetylglucosaminyltransferase, NEU: sialidase or neuraminidase, SLC: solute carrier, and ST: sialytransferase

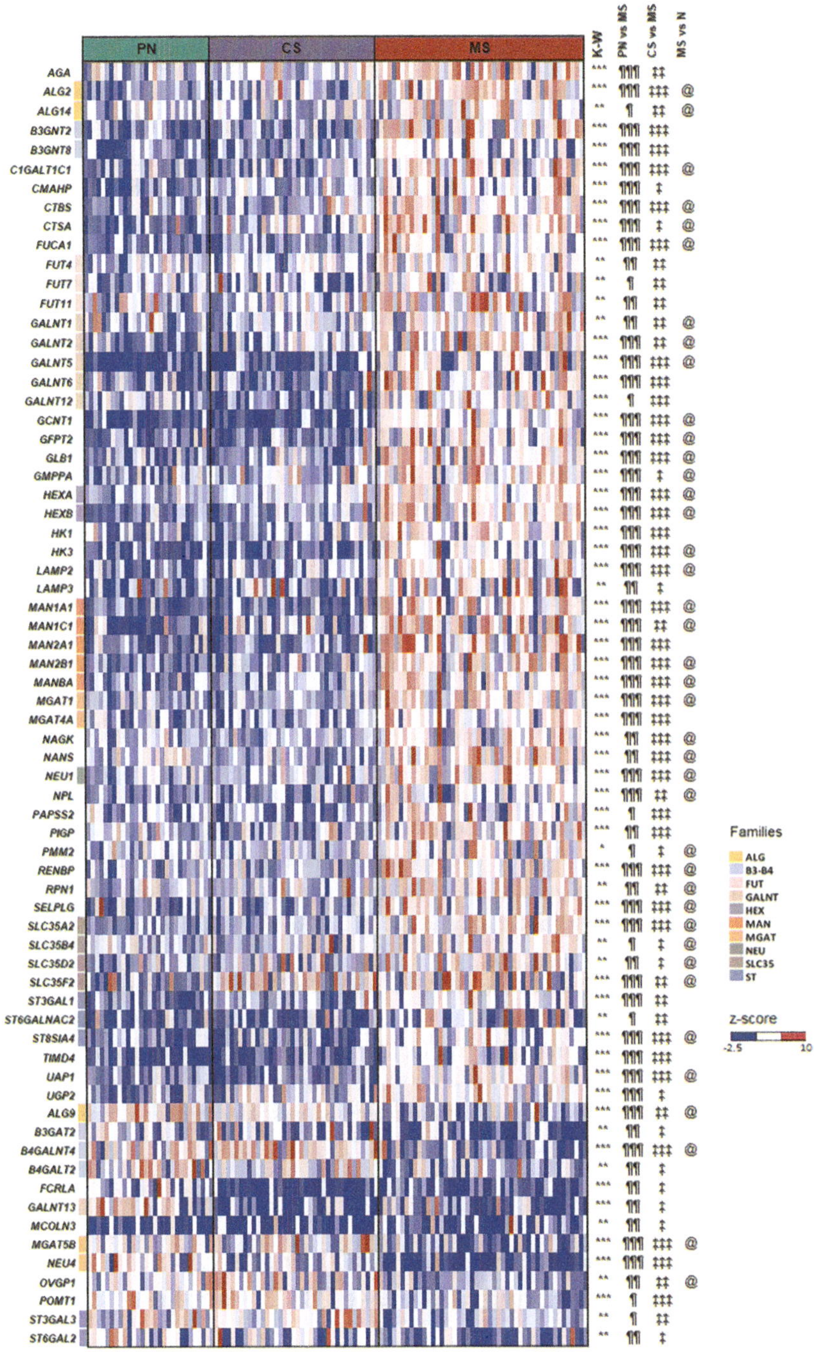

Fig. 4.2 Glycogenesexpression based on the glioblastoma TCGA-RNAseq dataset. The gene expression dataset was downloaded (Genomics Data Commons Data Portal - https://portal.gdc.cancer.gov/)

(continued)

4.2 Asparagine N-Linked Glycosylation Family (ALG)

The dolichol pathway is the first step of the N-glycosylation cascade in eukaryotes and consists of the assembly of a $Glc_3Man_9GlcNAc_2$ oligosaccharide linked to dolichol pyrophosphate at the ER membrane [56]. The linkage of each glycosyl residue is catalyzed by glycosyltransferases encoded by the ALG (asparagine-linked glycosylation) gene family.

Given the conserved nature of early stages of N-linked glycosylation, the function of ALG members was able to be established due to a variety of genetic studies in *Saccharomyces cerevisiae*. Mutations blocking specific stages of the biosynthetic pathway led to the characterization of their cellular functions in yeast and of each equivalent in higher eukaryotes [57, 58]. On the cytoplasmatic side of ER, the first step is catalyzed by Alg7p (also known as DPAGT1 in mammals), a GlcNAc-1-phosphotransferase [59]. This enzyme has been shown to be essential for cell viability and its activity can be inhibited by tunicamycin [60, 61]. A GlcNAc is attached, and five mannose residues are subsequently transferred from UDP-GlcNAc and GDP-Man, respectively. Each sugar residue is added by its respective enzyme (genes *ALG13/14, ALG1, ALG2,* and *ALG11*).

The $Man_5GlcNAc_2$-P-P-Dol molecule is flipped to the ER lumen following further extension by the addition of four mannoses and three glucose residues, transferred from Dol-P-Man and Dol-P-Glc, respectively, by the enzymes encoded by *ALG3*, *ALG9*, *ALG12*, *ALG6*, *ALG8*, and *ALG10*. Finally, the oligosaccharide is transferred to the Asn side chain within the N-X-S/Thr motif by oligosaccharyltransferase.

The correct function of ALG enzymes is essential to normal development. Glycosyltransferase gene variants have been associated with Congenital Disorders of Glycosylation (CGD) Type I, a set of rare diseases often involving the nervous system, but still showing great clinical heterogeneity [62, 63]. Most ALG variants have autosomal recessive inheritance, except for *ALG13-CDG* which is X-linked [63, 64].

Gene expression analysis of the GBM-TCGA-RNAseq dataset showed glycosyltransferase transcripts *ALG2* and *ALG14* were upregulated, while *ALG9* was downregulated in the mesenchymal GBM molecular subtype when compared to proneural and classical subtypes (Fig. 4.2; $p < 0.0001$, Kruskal-Wallis test, and $p < 0.001$ post-hoc Dunn test). Interestingly, all 3 transcripts presented higher expression than the non-neoplastic brain tissue (Figs. 4.2 and 4.3).

4.3 β-1,4-Galactosyltransferase (B4T) and β-1,3-Nacetylglucosaminyltransferase (B3T) Families

Glycosyltransferases are a large family consisting of 200–300 proteins involved in glycan biosynthesis [65]. The sequential action of these enzymes results in the formation of complex glycan structures that are present in prokaryotes and eukaryotes

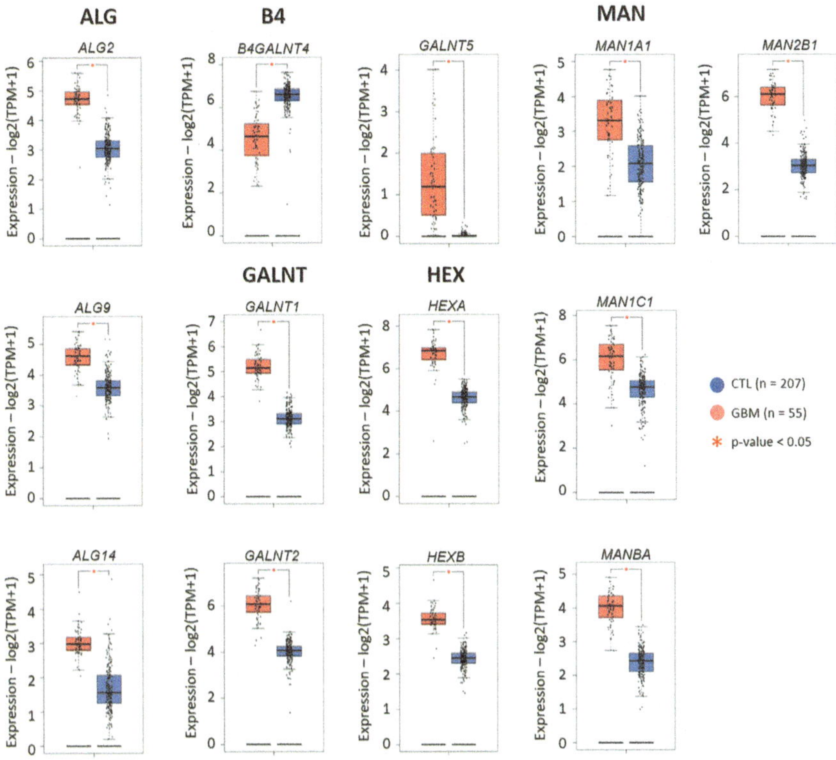

Fig. 4.3 Boxplots illustrating the expression profiles of genes coding for different families of enzymes in GBM mesenchymal molecular subtype and non-neoplastic brain tissue in the TCGA and GTEx datasets, respectively, compared by the GEPIA2 online tool (http://gepia2.cancer-pku.cn). The Y-axis represents the log2 (TPM + 1) of gene expression levels. The asterisk (*) indicates statistically significant differences ($p < 0.05$) between the two groups. GBM, glioblastoma; CTL, non-neoplastic brain tissue

[66]. Enzymes from the β-1,4-galactosyltransferases (B4T) and β-1,3-N acetylglucosaminyltransferase (B3T) subfamilies are key components in glycoprotein modification [67], responsible for elongating the polylactosamine of N- and O-glycans [68]. The enzymes alternate to add units of galactose (Gal) and N-acetylglucosamine (GlcNAc) by linking B1–4 and B1–3 to the growth chain [67, 68]. The B3T subfamily has eight members (B3GnT1–8), which were identified by a common family motif [65]. Among the protein members, polylactosamine synthesis activity was demonstrated to be strongest by N-acetyllactosaminide β-1,3-N-acetylglucosaminyltransferase 2 (B3GNT2) [69] and is mostly identified in cells of the immune system [70]. On the other hand, the B4T subfamily includes seven β 1,4-galactosyltransferases, two β-1,4-N-acetylgalactosaminyltransferases, and four chondroitin synthases that shared the WGXEDD/V/W motif [71].

The interaction between polylactosamine with lectins and other carbohydrate-binding proteins has been shown to trigger signal transductions and modulate different biological events, including proliferation and cell death [72]. Furthermore, the length of the polylactosamine chain influences the immune response. NK cells recognize cancer target cells in N glycans with very short polylactosamine chains [73]. On the other hand, long polylactosamine chains in the N-glycan are recognized to suppress excessive immune responses [74, 75]. Other events related to the B4T and B3T families include apoptosis, cell maturation, maintenance of immune tolerance, and modulation of immune cell function [75].

In our analysis, the differential regulation of *B3GAT2, B3GNT2, B3GNT8, B4GALNT4*, and *B4GALT2* transcripts was identified ($p < 0.05$, Kruskal-Wallis test, and $p < 0.01$ post-hoc Dunn test) among GBM subtypes. Furthermore, the *B4GALNT4* has been identified as downregulated in cancer compared to the healthy counterpart, Fig. 4.2. The B4T subfamily has important functions in breast cancer, which the upregulation of the *B4GALT1* promotes the proliferation and cell adhesion of cancer cells [76, 77]. Furthermore, overexpressions of *B4GALT1* and *B4GALT5* were observed in leukemia cells with acquired multidrug resistance mechanisms [71]. Dysregulation of B4T members has also been reported in lung cancer [78], melanoma [79], colorectal cancer [80–82], esophageal squamous cell carcinoma [83], and astrocytomas [26]. Interestingly, we identified *B4GALNT4* downregulation in the mesenchymal GBM subtype. Furthermore, the comparison between cancer tissue and healthy tissue also showed a downregulation of *B4GALNT4* in GBM, Fig. 4.3.

Proteins of the B3T subfamily are involved in the progression of several types of human cancer [84]. They have been reported to participate in tumor invasion and metastasis through the regulation of metalloproteinases and polylactosamine [85]. Increased *B3GNT8* expression has been identified in gliomas compared to non-neoplastic brain tissues (Fig. 4.3). Furthermore, *B3GNT8* levels increased with increasing pathological grade of gliomas. Genesilencing resulted in decreased polylactosamine and decreased cell proliferation, migration, and metastatic capacity *in vitro* and *in vivo* [86]. Overexpression of *B3GNT2* and *B3GNT8* in colorectal cancer resulted in increased potential for cancer cell invasion. On the other hand, gene knockdown resulted in the inhibition of invasive activity [85]. In triple-negative breast cancer, elevated levels of *B3GNT8* were related to a favorable outcome [87]. In gastric cancer, increased expression of *B3GNT8* has been correlated with lymph node metastases [88]. Here, *B3GNT2* and *B3GNT8* were increased in mesenchymal GBM, a subtype with a poor prognosis and greater potential for aggressiveness. Our analyses reinforce that the modulation of glycosyltransferases is correlated with cancer progression, suggesting that future studies can further evaluate and validate the role of this family in GBM.

4.4 Fucosyltransferase Family (FUT)

Fucosyltransferases, encoded by FUT genes, catalyze the transfer of fucose from GDP-fucose to their substrates, thus far encompassing thirteen genes. FUT1–11 catalyzes N-linked fucosylation at the Golgi apparatus, and POFUT1 and POFUT2 catalyze O-linked fucosylation at the ER. Recently a mitochondrial fucosyltransferase was described in Leishmania and *Trypanosoma brucei* parasites [89, 90]. FUT8 (α-1, 6-fucosyltransferase) is the only enzyme known to catalyze a core fucosylation, and the remaining FUTs participate in the terminal fucosylation machinery. Most of the secretory glycoproteins and membrane-bound glycoproteins on the cell surface undergo fucosylation. Fucosylation is involved in a variety of biological processes, including selectin-mediated leukocyte-endothelial adhesion, lymphocyte homing, ABO blood group histocompatibility, notch receptor signaling [91], embryo-fetal development, and host-microbe interactions [92]. Fucosylation, especially core fucosylation, is one of the most widely occurring cancer-associated changes in N-glycans [93].

The GBM-TCGA-RNAseq dataset showed significant upregulation of *FUT4, FUT7*, and *FUT11* in the mesenchymal subtype compared to proneural and classical subtypes ($p < 0.005$, Kruskal-Wallis test, and $p < 0.01$ post-hoc Dunn test for both comparisons mesenchymal *vs*. proneural and mesenchymal *vs*. classical; Fig. 4.2). However, no statistical differences were observed in GBM *vs*. non-neoplastic brain tissue.

FUT4 transfers fucose to N-acetyllactosamine polysaccharides to generate fucosylated carbohydrate structures, and it catalyzes the synthesis of the non-sialylated antigen, Lewis X (CD15). FUT4 was described as an oncoprotein in several tumor types. Recently, FUT4 was reported as a potential mediator of the tumor progression effect of the circularRNA poly(ADP-ribose) polymerase family member 4 (circPARP4)/miR-125a-5p pathway in glioma [94]. FUT4 has also been proposed as a stem-like cell marker in medulloblastoma [95] and GBM [96]. Similarly, *FUT4* downregulation by miR125a-5p was associated with a suppression of bladder cancer progression [97], and *FUT4* inhibition by miR26a and miR26b with decreased colon cancer aggressiveness [98]. FUT4 was also associated with multidrug resistance through activation of PI3K/Akt in hepatocarcinoma [99], with EMT by inactivation of GSK3β with translocation of NFkB to the nucleus in breast carcinoma [100], and with an oncogenic phenotype of acute myeloid leukemia stem cells by regulating fucosylated CD44 in Wnt/β-catenin pathway [101]. In multiple myeloma, homeobox B anti-sense1 (HOXB-AS1) recruited Elavi-like RNA binding protein 1 (ELAVL1), which stabilized FUT4 and led to tumor cell growth [102]. Interestingly, the lncRNA *HOXB-AS1* also promoted glioma cell proliferation, migration, and invasion through miR-885-3p/HOXB2 axis [103].

FUT7 is involved in the synthesis of the E-selectin-binding sialyl-Lewis X antigens (*CD15s*). CD15s, together with CD15, participates in the homing process of leukocytes and both are correlated with cancer progression and metastasis [104]. CD15 and CD15s are involved in the adhesion of cancer cells to brain endothelium through selectin E (*CD62E*) interaction [105]. In fact, the knockdown of FUT4/CD15 and FUT7/CD15s significantly reduced the adhesion of lung cancer cells and GBM cells to brain endothelium, suggesting that their upregulation may potentiate the transmigration of circulating cancer cells into the brain [106]. *FUT11* is presumed to encode for α(1,3)-fucosyltransferase due to the presence of the two main conserved motifs, but its activity has not been demonstrated yet in any species. While there is still no report concerning *FUT11* specifically in GBM, it was reported as highly expressed in gynecological cancer, predicting poor outcome [35], as well as in breast invasive ductal carcinoma [107] and hepatocarcinoma linked to the hypoxic environment [108]. Similarly, hypoxia-induced *FUT11* was associated with pancreatic carcinoma tumor growth and metastasis, through the stability of pyruvate dehydrogenase kinase 1 (PDK1) and activation of the AKT serine/threonine kinase/mechanistic target of rapamycin kinase (AKT/mTOR) signaling pathway [109]. Moreover, *FUT11* was considered a biomarker of clear cell renal cell carcinoma development [110].

Although *FUT8*expression level was not differentially expressed in the GBM-TCGA-RNAseq dataset among GBM molecular subtypes, *FUT8-AS1* overexpression was reported in a subgroup of GBM patients, associated with poor survival [111]. Another meta-analysis study found *FUT8* expression was associated with GBM patients below 40 years of age and with a shorter overall survival time of female patients [112]. *FUT8* upregulation was also described in patients under 10 years of age and associated with tumor recurrence in ependymomas. In medulloblastoma, *FUT8* expression levels have been associated with gender (male), tumor stage, wingless (WNT) and sonic hedgehog (SHH) molecular subgroups, and metastasis status [112]. Moreover, *FUT8* is frequently upregulated in liver, lung, colorectal, pancreas, prostate, breast, oral cavity, esophagus, and thyroid tumors, along with diffuse large B-cell lymphoma, while being downregulated in gastric cancer [25]. A recent study demonstrated stabilization and immunosuppression of B7H3 by FUT8 in triple-negative breast cancer. *B7H3* has also been described in neuroblastoma and GBM protecting tumor cells from natural killer mediated lysis [113, 114]. Therefore, inhibition of core fucosylation may be a therapeutic strategy to improve the immune response in GBM.

4.5 *N*-Acetylgalactosaminyltransferase Family (GALNT)

Aberrant mucin-typeO-glycosylation by glycosyltransferase, including the polypeptide N-acetylgalactosaminyltransferases (GALNTs), is a well-described hallmark in a variety of cancers. Members of the GALNT family function as the initiating enzymes that catalyze the transfer of a N-acetyl-D-galactosamine residue to a serine or threonine residue of the acceptor protein during O-linked oligosaccharide biosynthesis. The GALNT family is composed of 20 members, and in the GBM TCGA-RNAseq dataset analysis, five members (*GALNT1*, *GALNT2*, *GALNT5*, *GALNT6*, and *GALNT12*) were upregulated in the mesenchymal GBM molecular subtype when compared to proneural and classical subtypes ($p < 0.001$, Kruskal-Wallis test, and $p < 0.01$ post-hoc Dunn for both comparisons; Fig. 4.2). Among them, *GALNT1*, *GALNT2*, and *GALNT5* presented statistically higher expression levels than non-neoplastic brain tissue (Fig. 4.3). On the other hand, *GALNT13* was upregulated in proneural GBM molecular subtypes when compared to classical and mesenchymal subtypes.

Few studies are available of GALNT family expression analysis in gliomas. *GALNT2* was found to inhibit U251 cell migration by decreasing the expression of *MMP-9* and *TGF-β1* [19]. *GALNT2* expression levels were associated with an unfavorable prognosis and a higher tumor grade in human gliomas. GALNT2 facilitates the malignant characteristics of glioma by influencing EGFR O-glycosylation and phosphorylation and the subsequent downstream PI3K/Akt/mTOR axis [115]. Expression analysis and role of GALNTs in a broad spectrum of cancers were reviewed by [116, 117], and more recently by [118]. *GALNT6* and *GALNT14* are potential therapeutic targets.

4.6 Hexosaminidase Family (HEX)

Lysosomalhexosaminidases are enzymes that remove terminal N-acetyl-β-D-hexosamine from a number of neutral and anionic glycoconjugates [119]. Three different isoenzymes exist composed of two non-covalently linked heterodimeric subunits (α and β): Hex A (αβ), Hex B (ββ), and Hex S (αα). These two subunits are encoded by the *HEXA* and *HEXB* genes. Mutations in these genes have been associated with hereditary neurodegenerative diseases such as Tay-Sachs and Sandhoff disease [120, 121].

Elevated levels of hexosaminidases have been detected in colon, breast, gastric, and ovarian epithelial cancers [122–127]. In prostate cancer, downregulation of hexosaminidase was detected [128]. In particular, HEXA activity and expression were almost absent in PC3 and LNCaP cells. This downregulation was associated with hypermethylation of the *HEXA* promoter region. Moreover, overexpression of the α subunit in PC3 and LNCaP cells reduced sensitivity to ceramide treatment. Elevated hexosaminidase activity was detected in sera of patients with liver metastases compared with controls and cancer patients without metastasis [129]. In our data, *HEXA* and *HEXB* levels were found upregulated in the mesenchymal compared to the classic and proneural GBM subtypes ($p < 0.001$, Kruskal-Wallis test, and $p < 0.001$ post-hoc Dunn for both comparisons; Fig. 4.2). The expression in the mesenchymal subtype was also higher than non-neoplastic brain tissue (Fig. 4.3). Specific α-mannosidases and N-acetyl-β-hexosaminidase isoenzymes have been shown to generate paucimannose structures [130]. Paucimannose are glycan structures N-linked to proteins and consist of two N-acetylglucosamines, 1 to 3 mannose residues, and an option of a single fucose residue (mannose$_{1-3}$fucose$_{0-1}$N-acetylglucosamine$_2$). Initially described mainly in invertebrates and plants [131], their presence has also been described in vertebrates and humans, being associated with several (patho)physiological processes such as innate immunity, tumorigenesis, and cellular differentiation [132, 133]. In humans, proteins such as neutrophil cathepsin G, myeloperoxidase, elastase, azurocidin, and β-glucuronidase have been found as substrates for paucimannose structures [134–139]. Recently, a glycomic-centered survey of protein paucimannosylation of 34 cancer cell lines and 133 tissue samples spanning 11 cancer types against matching non-cancerous specimens revealed a wide expression of protein paucimannosylation. Colorectal, blood, and brain cancers had higher levels of paucimannosidic glycans which were also more abundant in tumors *versus* non-tumor tissue specimens. Interestingly, immunofluorescence of non-permeabilized U87MG and U138MG GBM cells with the monoclonal Mannitou antibody, specific for the trimannosidic paucimannose structure, showed that these epitopes are present on the cell surface [140, 141]. A172 and THP-1 staining with the same antibody revealed partial co-localization with HEXA [142]. These data indicate that elevated levels of paucimannosidic glycans in cancer cells and tissues are associated with differential expression of hexosaminidases and can influence the disease progression.

Carbohydrate-based pharmacological chaperones have been used to aid the folding of mutated hexosaminidases facilitating their transport to the lysosomes. N-acetyl glucosamine thiazoline, bisnaphthalimide, nitro-indan-1-one, and pyrrolo[3,4-d]pyridazin-1-one were identified as pharmacological chaperones that increased lysosomal HEX protein and activity levels in patient fibroblasts [143]. Interestingly, some of these compounds have already been used as anticancer agents [144–146]. Due to the high expression of hexosaminidases in the majority of the tumors, novel fluorescent probes such as HMRef-βGlcNAc were proposed as a cancer imaging tool and tested in cells, mouse model, and human colorectal cancer specimen [147].

4.7 Mannosidase Family (MAN)

Mannosidases (MANs) are a group of enzymes that remove mannose from polysaccharides and complex glycoconjugates composed of two types: α- and β-mannosidase. Members of the family of processing α-mannosidase are responsible for processing newly formed glycoproteins in the ER and Golgi into mature glycoproteins that contain highly complex and heterogeneous types of glycans. Lysosomal MAN is necessary for N-linked glycoprotein oligosaccharide catabolism.

Three genes coding for processing α-mannosidases, *MAN1A1*, *MAN1C1*, and *MAN2A1*, and two genes coding for lysosomal MAN, *MAN2B1* and *MANBA*, were upregulated in the mesenchymal GBM molecular subtype when compared to proneural and classical subtypes ($p < 0.005$, Kruskal-Wallis test, and $p < 0.01$ post-hoc Dunn test for both comparisons mesenchymal *vs.* proneural and mesenchymal *vs.* classical; Fig. 4.2). Furthermore, except for *MAN2A1*, all four genes presented significantly higher expression levels than a non-neoplastic brain tissue (Fig. 4.3). MAN1A1 (mannosidase alpha class 1A member 1) and its paralogue MAN1C1 (mannosidase alpha class 1C member 1) are involved in the maturation of N-linked oligosaccharides by trimming α(1,2)-linked mannose residues from $Man_9GlcNAc_2$. MAN2A1 (mannosidase alpha class 2A member 1) is localized in the Golgi and catalyzes the final step of N-glycan maturation.

Expression of these genes was not analyzed in gliomas. However, inhibitors of Golgi α-mannosidase II using functionalized pyrrolidine were able to decrease the growth of human GBM cells [148], indicating that the enzymes encoded by these genes are important for GBM, a hypothesis that is reinforced by studies in other types of cancer. The expression of the α-mannosidase coded by *MAN1A1* affected the adhesive properties of breast cancer cells and resulted in a poorer prognosis [33]. A similar unfavorable prognostic role of upregulated MANA1A was observed in ovarian cancer [149] and hepatocarcinoma [42]. High expression of MAN1C1 was described in non-Hodgkin lymphoma [150] and acute lymphoblastic leukemia [151]. However, expression of *MAN1C1* was decreased in hepatocarcinoma [42] and clear cell renal cell carcinoma cells [150], suggesting *MAN1C1* may be a tumor suppressor gene.

MAN2B1 (mannosidase alpha class 2B member 1) hydrolyzes terminal, non-reducing α-D-mannose residues in α-D-mannosides. MANBA (mannosidase beta) is the final exoglycosidase in the pathway for N-linked glycoprotein oligosaccharide catabolism, cleaving β-linked mannose residues. Lysosomal α-mannosidase activity was measured in different human gliomas, including astrocytomas and oligodendrogliomas; this activity was significantly higher in tumors than in non-neoplastic brain tissue, and the highest activity was observed in the higher malignant grade tumors, such as GBM [152]. Increased expression of *MAN2B1* was described in promyelocytic leukemia cell lines when compared with other leukemic cell lines [153]. Beyond activity, there are no studies to date focusing on *MANBA*expression in cancer.

4.8 N-Acetylglucosaminyltransferase Family (MGAT)

N-acetylglucosaminyltransferase, encoded by the *MGAT* gene, catalyzes the transfer of N-acetylglucosamine (GlcNAc) and regulates oligosaccharide branching on the newly synthesized glycoprotein in the Golgi apparatus. MGAT1 (α-1,3-mannosyl-glycoprotein 2-β-N-acetylglucosaminyltransferase) is a medial-Golgi enzyme and initiates complex N-linked carbohydrate formation, being essential for the conversion of high-mannose to hybrid and complex N-glycans [154]. MGAT2 (α-1,6-mannosyl-glycoprotein 2-β-N-acetylglucosaminyltransferase) catalyzes the conversion of oligomannose to complex N-glycans. MGAT3 (β-1,4-mannosyl-glycoprotein 4-β-N-acetylglucosaminyltransferase) transfers a GlcNAc residue to the β-linked mannose of the trimannosyl core of N-linked oligosaccharides and produces a bisecting GlcNAc. The introduction of bisecting GlcNAc precludes further branching catalyzed by N-acetylglucosaminyltransferase V (GnT-V, *MGAT5*). *MGAT4A* and *MGAT4B* (α-1,3-mannosyl glycoprotein 4-β-N-acetylglucosaminyltransferase) regulates the formation of tri- and multiantennary branching structures. MGTA5 (α-1,6-mannosylglycoprotein 6-β-N-acetylglucosaminyltransferase) catalyzes the addition of β-1,6-N-acetylglucosamine to the α-linked mannose of biantennary N-linked oligosaccharides present on the newly synthesized glycoproteins. *MGTA5B* (α-1,6-mannosylglycoprotein 6-β-N-acetylglucosaminyltransferase B), specifically expressed in the brain, is involved in the synthesis of complex cell surface N-glycans.

The GBM-TCGA-RNAseq dataset showed significant upregulation of *MGAT1* and *MGAT4A*, in contrast to a downregulation of *MGAT5B* in the mesenchymal subtype compared to proneural and classical subtypes ($p < 0.0001$, Kruskal-Wallis test, and $p < 0.0001$ post-hoc Dunn test for both comparisons mesenchymal *vs.* proneural and mesenchymal *vs.* classical; Fig. 4.2). The comparison of this family with non-neoplastic brain tissue revealed an upregulation of *MGAT1* and downregulation of *MGAT5B* in GBM (Fig. 4.4).

MGAT1 has been associated with glioma cell proliferation and migration by increasing the stability of the glucose transporter GLUT1 [150], which promotes cell proliferation and migration by regulating the EGFR/MAPK and integrin-β/Src/FAK pathways [155].

Although there is no report about *MGAT4A* in gliomas, it was described as downregulated in pancreatic cancer [156], and upregulated in choriocarcinoma [157]. A bioinformatic analysis of a microRNA study in hepatocarcinoma cell lines suggested that *let-7c*, a tumor-suppressor miRNA, might inhibit the metastatic ability of these cells inhibiting MGAT4A activity [158].

MGAT5B, also designated as GnT-IX, is a brain-specific enzyme regulated by defined chromatin modifiers, silenced by histone deacetylase (HDAC11), and activated by the O-GlcNAc transferase and ten-eleven translocation-3 complex [159]. There is no report about *MGAT5B* expression levels in brain tumors; however, it was upregulated in prostate cancer cell lines, and its products were predominantly detected in metastatic xenografts [160]. *MGAT5* (or MGAT5A), however, was described playing a key role in glioma stem cell migration speed via FAK signaling [161]. In cancer, *MGAT5* expression is regulated by the RAS-RAF-MAPK signaling pathway, and its upregulation results in loss of contact inhibition, cell motility, invasion, and metastasis [162].

In fact, the highly branched N-glycans catalyzed by *MGAT5* harbor higher affinities for galectin, resulting in a galectin lattice which modulates the turnover of integrin-ECM focal adhesion [163]. Additionally, the branched N-glycans attached to E-cadherin by *MGAT5* lead to incorrectly assembled and non-functional adherens junctions, impairing cell-cell adhesion and facilitating tumor cell motility [164]. ECM stiffness increases during tumor growth, and the thickness of glycocalyx and a specific glycosylation enzyme may regulate mechanotransduction and cell migration [161]. MGAT5 function is counteracted by MGAT3, which suppresses cancer metastasis [183] through EGFR, integrins, and cadherins [165], and suppresses tumor progression increasing the E-cadherin-mediated stability of adherens junctions. MGAT3 is also associated with suppression of EMT [97, 166].

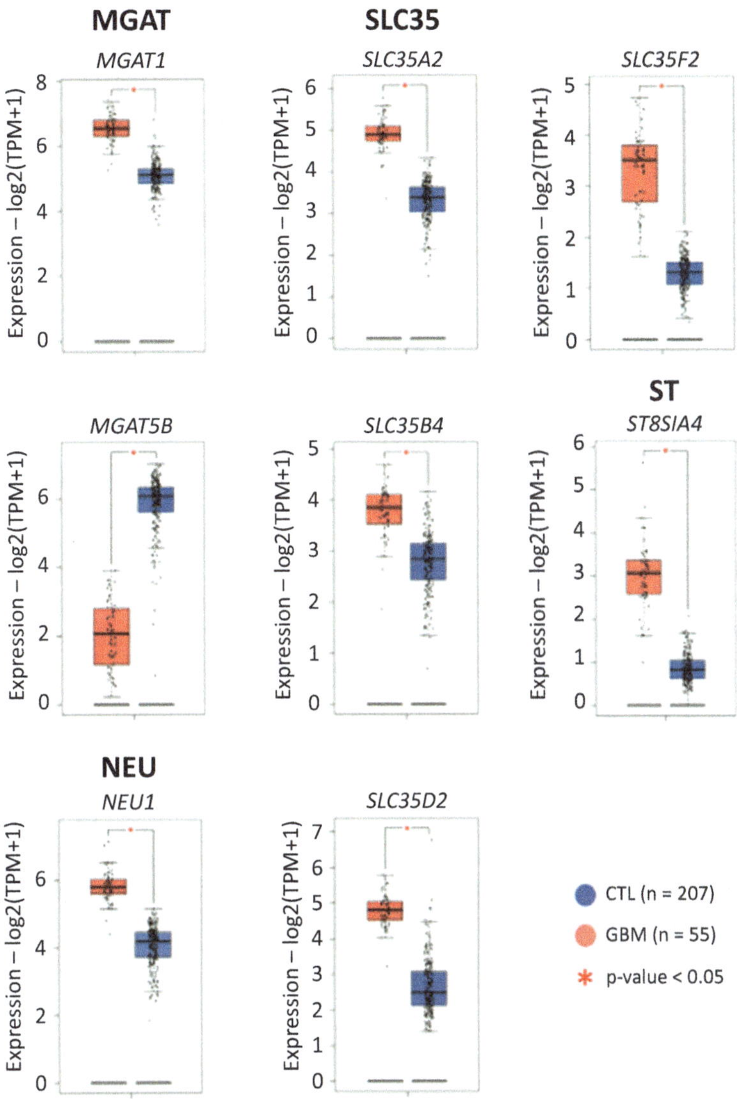

Fig. 4.4 Boxplots illustrating the expression profiles of genes coding for different families of enzymes in GBM mesenchymal molecular subtypes and non-neoplastic brain tissues in TCGA and GTEx, respectively, based on GEPIA2 analysis. The Y-axis represents the log2 (TPM + 1) of gene expression levels. An asterisk (*) indicates statistically significant differences ($p < 0.05$) between the two groups. GBM, glioblastoma; CTL, non-neoplastic brain tissue

4.9 Sialidase Family (NEU)

Sialidases, or neuraminidases (NEU), are glycosidases that catalyze the hydrolysis of α-glycosidic bonds linking sialic acid residues to carbohydrate groups of glycoproteins and glycolipids. Four types of sialidases have been described: *NEU1, NEU2, NEU3, and NEU4. NEU1* is involved in exocytosis, immune response, phagocytosis, and elastic fiber assembly; *NEU2* in myoblast and neuronal differentiation; and *NEU3* and *NEU4* in neuronal differentiation, apoptosis, and adhesion [167].

In gliomas, aberrant sialylation regulates important mechanisms of malignant cell behavior and associates with a worse prognosis. In fact, we observed a significant upregulation of *NEU1* in the mesenchymal GBM subtype ($p < 0.001$, Kruskal-Wallis test, and $p < 0.001$ post-hoc Dunn compared to proneural and classical subtypes; Figs. 4.2 and 4.4).

NEU1 has been described to desialylate the α-2,3-sialyl residues linked to β-galactosides of EGFR, and by removing steric hindrance of the receptor, it facilitates EGF binding with subsequent activation and downstream signaling [168]. NEU1 may also regulate extracellular and intracellular TOLL-like receptors in cancer cells through a cross-talk with matrix metalloproteinase-9 and G-protein coupled receptors, leading to tumor cell proliferation and immune evasion [169]. Moreover, NEU1 increased cell proliferation and migration in hepatocellular carcinoma [170].

NEU4 expression was downregulated in the mesenchymal GBM subtype in our analysis ($p < 0.001$, Kruskal-Wallis test, and $p < 0.001$ post-hoc Dunn compared to proneural and classical subtypes), but no statistical difference was observed when compared to non-neoplastic tissue (Figs. 4.2 and 4.4). However, *NEU4* hyperexpression was described in GBM stem cells, probably involving *OCT4* and *GLI1* upregulations, as demonstrated by inducing *NEU4*activation in more differentiated GBM cells [171].

4.10 Solute Carrier 35 Family (SLC35)

The solute carriers (SLC) are transporters of organic and inorganic molecules across cell or organelle membranes [172]. Currently, the family comprises 395 genes that are divided into 52 families [173]. Among their members, an estimated 31 are from the solute carrier 35 family (SLC35) [174], which is further subdivided into seven other subfamilies (SLC35A-G) [174]. Nucleotide sugar transporters (NSTs) belong to the SLC35 family and are hydrophobic proteins responsible for transporting nucleotide sugars and adenosine 3′-phospho 5′-phosphosulfate from the cytosol to the ER and/or the Golgi complex [172, 174].

NSTs form an essential link between the synthesis of nucleotide sugars and glycosylation in eukaryotic cells [175]. Additionally, they play a key role in regulating glycosylation and controlling the transport of activated sugars to glycosyltransferases [176]. Several members of the SLC35 family are still known as "orphan transporters", as their substrate molecules and their physiological functions are still poorly understood [174]. Nonetheless, the SLC35A2–5 subfamily, principally SLC35A2 and SLC35A3, are related to the galactosylation of N-glycans [177]. SLC35F2 plays an important role in the delivery of anticancer drugs, highlighting the role of this family in health and disease. In addition, the family of NSTs can also be targeted to reduce cancer cell hypersialylation, which is recognized to increase tumor survival and proliferation [178, 179].

Our analyses identified four members of the SLC35 family that were differentially regulated among the GBM subtypes: *SLC35A2, SLC35F2* ($p < 0.001$, Kruskal-Wallis test, and $p < 0.01$ post-hoc Dunn), *SLC35B4,* and *SLC35D2* ($p < 0.01$, Kruskal-Wallis test, and $p < 0.05$ post-hoc Dunn) (Fig. 4.2). *SLC35A2*, *SLC35B4*, *SLC35D2*, and *SLC35F2* were upregulated in GBM compared to non-neoplastic brain tissues (Fig. 4.4). Previous studies have shown that SLC35F2 plays a key role in the biological processes of different types of cancer [103, 180]. In bladder cancer, increased SLC35F2 levels were associated with increased tumor invasiveness in *in vitro* models, which was confirmed by its knockdown in a murine model that reduced tumorigenesis [102]. The invasive effects of *SLC35F2* were also observed by its upregulation in human muscle-invasive bladder cancer compared to non-muscle invasive cancer [180].

The overexpression of *SLC35F2* was also detected in non-small cell lung cancer tissues compared to normal and adjacent tissues [181]. In prostate cancer, the single nucleotide polymorphism (SNPs) rs1646724 has been identified to affect the expression of the *SLC35B4* gene. Furthermore, it has been shown that a downregulation of *SLC35B4* suppresses the proliferation, migration, and invasion of cancer cells [182]. To date, there are no results in the literature on the role of the SLC35 family in GBM. Based on this, we draw attention to our *in silico* results that shed light on the importance of the roles of this family in the genesis and progression of GBM subtypes.

4.11 Sialyltransferase Family (ST)

Sialyltransferases (STs) are a family of enzymes involved in the biosynthesis of sialic acid-modified glycoproteins and glycolipids [183, 184], which are negatively charged monosaccharides commonly linked to terminal residues in glycans [185, 186]. The mechanisms of action are via transfer of a sialic acid residue from its activated sugar nucleotide donor cytidine 5′-monophosphate to a variety of acceptor molecules, which include structures terminated with Gal, GalNAc, or another sialic acid moiety [185–188]. STs catalyze the formation of a glycosidic bond between the C2 of the donor sialic acid and a C3, C6, or C8 hydroxyl of a glycan acceptor, thereby being classified into four main subfamilies based on the bonds the enzymes form: β-galactoside α-2,3-sialyltransferase (ST3Gal), β-galactoside α-2,6- sialyltransferase (ST6Gal), GalNAc α-2, 6-sialyltransferase (ST6GalNAc), and α-2,8-sialyltransferase (ST8SIA). Currently, it is estimated that the family of STs is composed of 20 proteins, 15 of which were cloned from human sources [189].

An overexpression of STs has been described in approximately 40–60% of tumor cell surfaces [39]. Furthermore, records demonstrate that hypersialylation is correlated with metastatic events, cancer cell survival, increased invasion, and migration of different cancers, including breast, ovarian, prostate, and pancreatic cancers [187]. Sialylation levels are additionally related to immune evasion, less effective treatment against cancer, and blocking of vital pathway signaling [190]. Although increased sialylation is related to a worse prognosis in different types of cancer, the effects on neural tissues are not known thus far, since polysialylation is an important event in the regulation and regeneration of neural systems [179].

Our analyses identified ST modulations, with upregulation of *ST3GAL1, ST8SIA4* ($p < 0.001$, Kruskal-Wallis test, and $p < 0.01$ post-hoc Dunn compared to proneural and classical subtypes), *ST6GALNAC2* ($p = 0.004$, Kruskal-Wallis test, and $p < 0.05$ post-hoc Dunn compared to proneural and classical subtypes), and downregulation of *ST3GAL3 and ST6GAL2* in the mesenchymal subtype of GBM in relation to classical and proneural subtypes ($p = 0.001$, Kruskal-Wallis test, and $p < 0.05$ post-hoc Dunn for both comparisons), (Fig. 4.2); *ST8SIA4* was significantly upregulated in GBM compared to non-neoplastic brain tissue (Fig. 4.4).

Results from *in vivo* and *in vitro* studies have shown that *ST3GAL1* leads to metastasis in melanoma. However, a knockout of the *ST3GAL1* gene reduced the metastatic behavior of cells [191]. In breast cancer, ST3GAL1-mediated O-linked sialylation of CD55 has been shown to promote cancer cell evasion of the immune response [192]. Analyses using MALDI-TOF MS showed that the overexpressed ST8SIA4 protein is also associated with a poor prognosis in breast cancer [193].

In contrast to the GBM results demonstrated here, *ST3GAL3* was identified overexpressed in the metastatic ovarian cell line HO8910PM compared to the SKOV3 cell line [151]. Furthermore, the same behavior has been identified in pancreatic ductal adenocarcinoma, in which a downregulation of *ST3GAL3* resulted in decreased invasiveness and migration of cancer cells [194]. Based on these findings, we emphasize that the investigation of the role of STs in the genesis and progression of GBM is of fundamental importance. Furthermore, there may also be a unique molecular signature in this type of cancer, as STs play important roles in metastatic events, as well as in the regulation and regeneration of neural systems.

4.12 Conclusions and Outlook

Our study reveals altered mRNA expression of genes coding for proteins involved in the protein glycosylation biosynthesis in GBM molecular subtypes. The majority of them were upregulated in the mesenchymal subtype, the most aggressive and malignant subtype, indicating a glycosylation signature involved in the EMT process. Most of these genes were also dysregulated in comparison with non-neoplastic

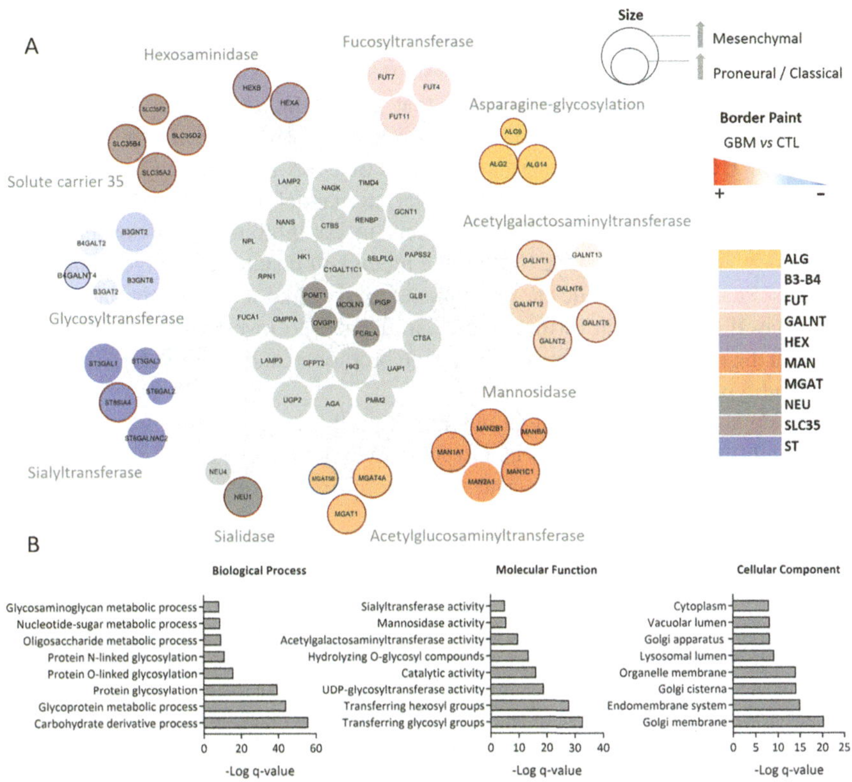

Fig. 4.5 Glycogene interaction network obtained from the STRING database analysis (https://string-db.org/), using the following parameters: medium confidence score (0.400); textmining, co-expression, and neighborhood enabled. Border colors indicate upregulated (red) and downregulated (blue) genes in glioblastoma (GBM) compared to healthy counterparts. Genes with white borders did not show significant differential expression (p-value <0.05). The size of glycogenes indicates regulation between GBM subtypes. Larger and smaller sizes indicate upregulation and downregulation in the mesenchymal subtype compared to the classic and proneural subtypes, respectively (**a**). Top 8 gene ontology terms related to the genes shown in the interaction network. All biological processes, molecular functions, and cellular components presented have a q-value <0.05 (Benjamini-Hochberg) (**b**)

brain tissues. In particular, a protein-protein interaction map of the differentially regulated genes revealed an interconnected network where specific families previously discussed were highlighted (Fig. 4.5a). Gene ontology analyses revealed an enrichment of biological processes associated with nucleotide metabolism; protein N- and O-linked glycosylation; molecular functions associated with sialyltransferases, mannosidases, and acetylgalactomanimyltransferases; and cellular components associated with ER, Golgi, and lysosomes (Fig. 4.5b). Aberrant expression of glycogenes in cancer has been linked to differential methylation of DNA promoters [195]. Indeed, an *in silico* analysis of the expression and methylation states of 86 glycogenes in melanoma, hepatocellular, breast, and cervical cancers showed that ten glycogenes (*GALNT3*, *GALNT6*, *GALNT7*, *GALNT14*, *MGAT3*, *MAN1A1*, *MAN1C1*, *ST3GAL2*, *ST6GAL1*, and *ST8SIA3*) are modulated on the two levels. In particular, *MGAT5B* was shown to be dysregulated and differentially methylated in different cancers. These data showed that epigenetic regulation of glycogenes impacts the aberrant glycan expression on cancer cells dictating their behavior. Moreover, HeLa cells treated with inhibitors of DNA methylation (zebularin and 5-aza-2′-deoxycytidine) and histone deacetylation (trichostatin A and Na-butyrate) showed a modulation of N-linked glycans in an inhibitor- and concentration-dependent manner [196]. Due to that, the methylation status of a glycogene could serve as a cancer diagnostic or a prognostic marker and a therapeutic target [197, 198]. Based on our data, further studies are needed to investigate the correlations between the methylation status of glycogene promoters and their expression level in GBM subtypes.

The data presented here highlight the importance of protein glycosylation in GBM and the huge potential in identifying novel therapeutic and diagnostic markers. However, further validation is needed to confirm their role in GBM during EMT as cancer targets.

Acknowledgments We are grateful for the financial support provided by the São Paulo Research Foundation (FAPESP; grant numbers 2018/18257-1 (GP), 2018/15549-1 (GP), 2021/00140-3 (JMDS), 2001/12898-4 (SKNM), 2004/12133-6 (SKNM), 2013/02162-8 (SKNM), 2014/50137-5 (SKNM), 2020/02988-7 (SKNM, SMO-S)); the National Council for Scientific and Technological Development (CNPq; grant number 305730/2015-0 (SKNM) and "Bolsa de Produtividade" (SKNM and GP)); FMUSP (SKNM and SMO-S); and the Coordination of Superior Level Staff Improvement (CAPES; grant numbers NUFFIC 062/15 (SKNM) and 9999.001625/2015-02 (SKNM) and a PNPD stipend (LRF)).

References

1. Reily, C., Stewart, T. J., Renfrow, M. B., & Novak, J. (2019). Glycosylation in health and disease. *Nature Reviews. Nephrology, 15*, 346–366.
2. Taniguchi, N., Honke, K., Fukuda, M., & Clausen, H. (2002). *Handbook of glycosyltransferases and related genes*. Springer.
3. Narimatsu, H. (2006). Human glycogene cloning: Focus on beta 3-glycosyltransferase and beta 4-glycosyltransferase families. *Current Opinion in Structural Biology, 16*, 567–575.
4. Pinho, S. S., & Reis, C. A. (2015). Glycosylation in cancer: Mechanisms and clinical implications. *Nature Reviews. Cancer, 15*, 540–555.
5. Kamei, N., et al. (2010). Definitive evidence that a single N-glycan among three glycans on inducible costimulator is required for proper protein trafficking and ligand binding. *Biochemical and Biophysical Research Communications, 391*, 557–563.
6. Arigoni-Affolter, I., et al. (2019). Mechanistic reconstruction of glycoprotein secretion through monitoring of intracellular N-glycan processing. *Science Advances, 5*, eaax8930.
7. Siegel, R. L., Miller, K. D., & Jemal, A. (2019). Cancer statistics, 2019. *CA: a Cancer Journal for Clinicians, 69*, 7–34.
8. Dolecek, T. A., Propp, J. M., Stroup, N. E., & Kruchko, C. (2012). CBTRUS statistical report: Primary brain and central nervous system tumors diagnosed in the United States in 2005-2009. *Neuro-Oncology, 14*(Suppl 5), 1–49.

9. Molinaro, A. M., Taylor, J. W., Wiencke, J. K., & Wrensch, M. R. (2019). Genetic and molecular epidemiology of adult diffuse glioma. *Nature Reviews. Neurology, 15*, 405–417.
10. Sanai, N., Alvarez-Buylla, A., & Berger, M. S. (2005). Neural stem cells and the origin of gliomas. *The New England Journal of Medicine, 353*, 811–822.
11. Hanif, F., Muzaffar, K., Perveen, K., Malhi, S. M., & Simjee, S. U. (2017). Glioblastoma Multiforme: A review of its epidemiology and pathogenesis through clinical presentation and treatment. *The Asian Pacific Journal of Cancer Prevention (APJCP), 18*, 3–9.
12. Dusoswa, S., et al. (2018). TMIC-28. Glioblastoma exploits cell surface glycosylation-mediated immune regulatory circuits for immune escape. *Neuro-Oncology, 20*, vi262.
13. Oronsky, B., Reid, T. R., Oronsky, A., Sandhu, N., & Knox, S. J. (2020). A review of newly diagnosed glioblastoma. *Frontiers in Oncology, 10*, 574012.
14. Wang, Q., et al. (2017). Tumor evolution of glioma-intrinsic gene expression subtypes associates with immunological changes in the microenvironment. *Cancer Cell, 32*, 42–56.e6.
15. Verhaak, R. G. W., et al. (2010). Integrated genomic analysis identifies clinically relevant subtypes of glioblastoma characterized by abnormalities in PDGFRA, IDH1, EGFR, and NF1. *Cancer Cell, 17*, 98–110.
16. Yan, H., et al. (2009). IDH1 and IDH2 mutations in gliomas. *The New England Journal of Medicine, 360*, 765–773.
17. Sidaway, P. (2017). CNS cancer: Glioblastoma subtypes revisited. *Nature Reviews. Clinical Oncology, 14*, 587.
18. Wahl, D. R., & Lawrence, T. S. (2019). No sugar added: A new strategy to inhibit glioblastoma receptor tyrosine kinases. *Clinical Cancer Research : An Official Journal of the American Association for Cancer Research, 25*, 455–456.
19. Liu, J., Yang, L., Jin, M., Xu, L., & Wu, S. (2011). Regulation of the invasion and metastasis of human glioma cells by polypeptide N-acetylgalactosaminyltransferase 2. *Molecular Medicine Reports, 4*, 1299–1305.
20. Yen, H.-Y., et al. (2015). Effect of sialylation on EGFR phosphorylation and resistance to tyrosine kinase inhibition. *Proceedings of the National Academy of Sciences of the United States of America, 112*, 6955–6960.
21. Whitson, K. B., et al. (2005). Functional effects of glycosylation at Asn-579 of the epidermal growth factor receptor. *Biochemistry, 44*, 14920–14931.
22. Kaszuba, K., et al. (2015). N-glycosylation as determinant of epidermal growth factor receptor conformation in membranes. *Proceedings of the National Academy of Sciences of the United States of America, 112*, 4334–4339.
23. Matsumoto, K., et al. (2008). N-glycan fucosylation of epidermal growth factor receptor modulates receptor activity and sensitivity to epidermal growth factor receptor tyrosine kinase inhibitor. *Cancer Science, 99*, 1611–1617.
24. Liwosz, A., Lei, T., & Kukuruzinska, M. A. (2006). N-glycosylation affects the molecular organization and stability of E-cadherin junctions. *The Journal of Biological Chemistry, 281*, 23138–23149.
25. Liao, C., et al. (2021). FUT8 and protein Core Fucosylation in Tumours: From diagnosis to treatment. *Journal of Cancer, 12*, 4109–4120.
26. Xu, S., et al. (2001). Over-expression of beta-1,4-galactosyltransferase I, II, and V in human astrocytoma. *Journal of Cancer Research and Clinical Oncology, 127*, 502–506.
27. Xu, Y., et al. (2017). N-glycosylation at Asn 402 stabilizes N-cadherin and promotes cell-cell adhesion of glioma cells. *Journal of Cellular Biochemistry, 118*, 1423–1431.
28. Gu, J., & Taniguchi, N. (2004). Regulation of integrin functions by N-glycans. *Glycoconjugate Journal, 21*, 9–15.
29. Marsico, G., Russo, L., Quondamatteo, F., & Pandit, A. (2018). Glycosylation and integrin regulation in cancer. *Trends Cancer, 4*, 537–552.
30. Zhao, Y., et al. (2006). Deletion of core fucosylation on alpha3beta1 integrin down-regulates its functions. *The Journal of Biological Chemistry, 281*, 38343–38350.

31. Suzuki, M., et al. (2005). Polysialic acid facilitates tumor invasion by glioma cells. *Glycobiology, 15*, 887–894.
32. Vajaria, B. N., Patel, K. R., Begum, R., & Patel, P. S. (2016). Sialylation: An avenue to target cancer cells. *Pathology & Oncology Research POR, 22*, 443–447.
33. Legler, K., et al. (2018). Reduced mannosidase MAN1A1 expression leads to aberrant N-glycosylation and impaired survival in breast cancer. *British Journal of Cancer, 118*, 847–856.
34. Khan, T., & Cabral, H. (2021). Abnormal glycosylation of cancer stem cells and targeting strategies. *Frontiers in Oncology, 11*, 649338.
35. Zhang, C., et al. (2017). TRIM8 regulates stemness in glioblastoma through PIAS3-STAT3. *Molecular Oncology, 11*, 280–294.
36. Yang, H.-F., et al. (2017). Fentanyl promotes breast cancer cell Stemness and epithelial-mesenchymal transition by upregulating α1, 6-Fucosylation via Wnt/β-catenin signaling pathway. *Frontiers in Physiology, 8*, 510.
37. Zhou, F., et al. (2010). Alpha2,3-Sialylation regulates the stability of stem cell marker CD133. *Journal of Biochemistry (Tokyo), 148*, 273–280.
38. Tao, C., et al. (2020). Genomics and prognosis analysis of epithelial-mesenchymal transition in glioma. *Frontiers in Oncology, 10*, 183.
39. Pucci, M., Malagolini, N., & Dall'Olio, F. (2021). Glycobiology of the epithelial to mesenchymal transition. *Biomedicine, 9*, 770.
40. Lange, T., Samatov, T. R., Tonevitsky, A. G., & Schumacher, U. (2014). Importance of altered glycoprotein-bound N- and O-glycans for epithelial-to-mesenchymal transition and adhesion of cancer cells. *Carbohydrate Research, 389*, 39–45.
41. Maupin, K. A., et al. (2010). Glycogene expression alterations associated with pancreatic cancer epithelial-mesenchymal transition in complementary model systems. *PLoS One, 5*, e13002.
42. Tu, H.-C., et al. (2017). Up-regulation of golgi α-mannosidase IA and down-regulation of golgi α-mannosidase IC activates unfolded protein response during hepatocarcinogenesis. *Hepatology Communications, 1*, 230–247.
43. Hassani, Z., et al. (2017). Phostine PST3.1a targets MGAT5 and inhibits glioblastoma-initiating cell invasiveness and proliferation. *Molecular Cancer Research, 15*, 1376–1387.
44. Zhang, N., et al. (2020). Loss of core fucosylation enhances the anticancer activity of cytotoxic T lymphocytes by increasing PD-1 degradation. *European Journal of Immunology, 50*, 1820–1833.
45. Okada, M., et al. (2017). Blockage of Core Fucosylation reduces cell-surface expression of PD-1 and promotes anti-tumor immune responses of T cells. *Cell Reports, 20*, 1017–1028.
46. Dusoswa, S. A., et al. (2019). Glycan modification of glioblastoma-derived extracellular vesicles enhances receptor-mediated targeting of dendritic cells. *Journal of Extracellular Vesicles, 8*, 1648995.
47. Läubli, H., et al. (2021). Tools to study and target the Siglec-sialic acid axis in cancer. *The FEBS Journal, 288*, 6206–6225.
48. Angata, T., et al. (2002). Cloning and characterization of human Siglec-11. A recently evolved signaling molecule that can interact with SHP-1 and SHP-2 and is expressed by tissue macrophages, including brain microglia. *The Journal of Biological Chemistry, 277*, 24466–24474.
49. Zhou, Y., et al. (2017). Inhibition of fucosylation by 2-fluorofucose suppresses human liver cancer HepG2 cell proliferation and migration as well as tumor formation. *Scientific Reports, 7*, 11563.
50. Huang, J., et al. (2021). The noncanonical role of EZH2 in cancer. *Cancer Science, 112*, 1376–1382.
51. Baro, M., Lopez Sambrooks, C., Quijano, A., Saltzman, W. M., & Contessa, J. (2019). Oligosaccharyltransferase inhibition reduces receptor tyrosine kinase activation and enhances glioma Radiosensitivity. *Clinical Cancer Research : An Official Journal of the American Association for Cancer Research, 25*, 784–795.

52. Shields, R. L., et al. (2002). Lack of fucose on human IgG1 N-linked oligosaccharide improves binding to human Fcgamma RIII and antibody-dependent cellular toxicity. *The Journal of Biological Chemistry, 277*, 26733–26740.
53. Kuo, Y.-C., Wang, L.-J., & Rajesh, R. (2019). Targeting human brain cancer stem cells by curcumin-loaded nanoparticles grafted with anti-aldehyde dehydrogenase and sialic acid: Colocalization of ALDH and CD44. *Materials Science & Engineering. C, Materials for Biological Applications, 102*, 362–372.
54. Xu, B., et al. (2020). Selenium nanoparticles reduce glucose metabolism and promote apoptosis of glioma cells through reactive oxygen species-dependent manner. *Neuroreport, 31*, 226–234.
55. Munkley, J., & Scott, E. (2019). Targeting aberrant sialylation to treat cancer. *Medicine Basel Switzerland, 6*, E102.
56. Burda, P., & Aebi, M. (1999). The dolichol pathway of N-linked glycosylation. *Biochimica et Biophysica Acta, 1426*, 239–257.
57. Herscovics, A., & Orlean, P. (1993). Glycoprotein biosynthesis in yeast. *FASEB J. off. Publ. Fed. Am. Soc. Experimental Biology, 7*, 540–550.
58. Kukuruzinska, M. A., Bergh, M. L., & Jackson, B. J. (1987). Protein glycosylation in yeast. *Annual Review of Biochemistry, 56*, 915–944.
59. Eckert, V., et al. (1998). Cloning and functional expression of the human GlcNAc-1-P transferase, the enzyme for the committed step of the dolichol cycle, by heterologous complementation in Saccharomyces cerevisiae. *Glycobiology, 8*, 77–85.
60. Yoo, J., et al. (2018). GlcNAc-1-P-transferase-tunicamycin complex structure reveals basis for inhibition of N-glycosylation. *Nature Structural & Molecular Biology, 25*, 217–224.
61. Barnes, G., Hansen, W. J., Holcomb, C. L., & Rine, J. (1984). Asparagine-linked glycosylation in Saccharomyces cerevisiae: Genetic analysis of an early step. *Molecular and Cellular Biology, 4*, 2381–2388.
62. Bieberich, E. (2014). Synthesis, processing, and function of N-glycans in N-glycoproteins. *Advance Neurobiology, 9*, 47–70.
63. Francisco, R., et al. (2019). The challenge of CDG diagnosis. *Molecular Genetics and Metabolism, 126*, 1–5.
64. Alsharhan, H., et al. (2021). ALG13 X-linked intellectual disability: New variants, glycosylation analysis, and expanded phenotypes. *Journal of Inherited Metabolic Disease, 44*, 1001–1012.
65. Taniguchi, N., & Korekane, H. (2011). Branched N-glycans and their implications for cell adhesion, signaling and clinical applications for cancer biomarkers and in therapeutics. *BMB Reports, 44*, 772–781.
66. Wassler, M. (2013). β1,4-galactosyltransferases, potential modifiers of stem cell pluripotency and differentiation. *Pluripotent Stem Cells (IntechOpen)*. https://doi.org/10.5772/54376
67. Manabe, S., Ueki, A., & Ito, Y. (2008). Reductive deprotection of propargyl ether by a SmI2–amine–water system and its application to polymer-supported oligosaccharide synthesis. *Tetrahedron Letters, 49*, 5159–5161.
68. Isaji, T., et al. (2010). Functional roles of the bisecting GlcNAc in integrin-mediated cell adhesion. *Methods in Enzymology, 480*, 445–459.
69. Togayachi, A., Sato, T., & Narimatsu, H. (2006). Comprehensive enzymatic characterization of glycosyltransferases with a beta3GT or beta4GT motif. *Methods in Enzymology, 416*, 91–102.
70. Hao, Y., et al. (2021). Structures and mechanism of human glycosyltransferase β1,3-N-acetylglucosaminyltransferase 2 (B3GNT2), an important player in immune homeostasis. *The Journal of Biological Chemistry, 296*, 100042.
71. Zhou, H., et al. (2013). B4GALT family mediates the multidrug resistance of human leukemia cells by regulating the hedgehog pathway and the expression of p-glycoprotein and multidrug resistance-associated protein 1. *Cell Death & Disease, 4*, e654.

72. Brewer, C. F., Miceli, M. C., & Baum, L. G. (2002). Clusters, bundles, arrays and lattices: Novel mechanisms for lectin-saccharide-mediated cellular interactions. *Current Opinion in Structural Biology, 12*, 616–623.
73. Ohyama, C., Tsuboi, S., & Fukuda, M. (1999). Dual roles of sialyl Lewis X oligosaccharides in tumor metastasis and rejection by natural killer cells. *The EMBO Journal, 18*, 1516–1525.
74. Togayachi, A., et al. (2007). Polylactosamine on glycoproteins influences basal levels of lymphocyte and macrophage activation. *Proceedings of the National Academy of Sciences of the United States of America, 104*, 15829–15834.
75. Togayachi, A., & Narimatsu, H. (2014). UDP-gal: BetaGlcNAc Beta 1,3-galactosyltransferase, polypeptide 5 (B3GALT5). In N. Taniguchi et al. (Eds.), *Handbook of glycosyltransferases and related genes* (pp. 89–99). Springer. https://doi.org/10.1007/978-4-431-54240-7_88
76. Villegas-Comonfort, S., Serna-Marquez, N., Galindo-Hernandez, O., Navarro-Tito, N., & Salazar, E. P. (2012). Arachidonic acid induces an increase of β-1,4-galactosyltransferase I expression in MDA-MB-231 breast cancer cells. *Journal of Cellular Biochemistry, 113*, 3330–3341.
77. Li, M., Song, L., & Qin, X. (2010). Glycan changes: Cancer metastasis and anti-cancer vaccines. *Journal of Biosciences, 35*, 665–673.
78. Zhu, X., et al. (2005). Elevated beta1,4-galactosyltransferase I in highly metastatic human lung cancer cells. Identification of E1AF as important transcription activator. *The Journal of Biological Chemistry, 280*, 12503–12516.
79. Johnson, F. M., & Shur, B. D. (1999). The level of cell surface beta1,4-galactosyltransferase I influences the invasive potential of murine melanoma cells. *Journal of Cell Science, 112*(Pt 16), 2785–2795.
80. Chen, W.-S., Chang, H.-Y., Li, C.-P., Liu, J. M., & Huang, T.-S. (2005). Tumor beta-1,4-galactosyltransferase IV overexpression is closely associated with colorectal cancer metastasis and poor prognosis. *Clinical Cancer Research : An Official Journal of the American Association for Cancer Research, 11*, 8615–8622.
81. Che, M.-I., et al. (2014). β1, 4-N-acetylgalactosaminyltransferase III modulates cancer stemness through EGFR signaling pathway in colon cancer cells. *Oncotarget, 5*, 3673–3684.
82. Ichikawa, T., et al. (1999). Expression of N-acetyllactosamine and beta1,4-galactosyltransferase (beta4GalT-I) during adenoma-carcinoma sequence in the human colorectum. *Journal of Histochemistry and Cytochemistry: Official Journal of Histochemistry Society, 47*, 1593–1602.
83. Baba, H., et al. (2020). Expression and malignant potential of B4GALNT4 in esophageal squamous cell carcinoma. *Annals of Surgical Oncology, 27*, 3247–3256.
84. Zhou, D., et al. (1999). A beta-1,3-N-acetylglucosaminyltransferase with poly-N-acetyllactosamine synthase activity is structurally related to beta-1,3-galactosyltransferases. *Proceedings of the National Academy of Sciences of the United States of America, 96*, 406–411.
85. Jiang, Z., et al. (2018). β3GnT8 promotes colorectal cancer cells invasion via CD147/MMP2/Galectin3 Axis. *Frontiers in Physiology, 9*, 588.
86. Liu, J., et al. (2014). High expression of β3GnT8 is associated with the metastatic potential of human glioma. *International Journal of Molecular Medicine, 33*, 1459–1468.
87. Okazaki, M., et al. (2021). Biological and Clinicopathological implications of Beta-3-N-acetylglucosaminyltransferase 8 in triple-negative breast cancer. *Anticancer Research, 41*, 845–858.
88. Shen, L., Dong, X., Yu, M., Luo, Z., & Wu, S. (2017). β3GnT8 promotes gastric cancer invasion by regulating the glycosylation of CD147. *Journal of Cancer, 8*, 314–322.
89. Guo, H., et al. (2021). A broadly active fucosyltransferase LmjFUT1 whose mitochondrial localization and activity are essential in parasitic Leishmania. *Proceedings of the National Academy of Sciences of the United States of America, 118*, e2108963118.

90. Bandini, G., et al. (2021). An essential, kinetoplastid-specific GDP-Fuc: β-D-Gal α-1,2-fucosyltransferase is located in the mitochondrion of Trypanosoma brucei. *eLife, 10*, e70272.
91. Okajima, T., Xu, A., Lei, L., & Irvine, K. D. (2005). Chaperone activity of protein O-fucosyltransferase 1 promotes notch receptor folding. *Science, 307*, 1599–1603.
92. Becker, D. J., & Lowe, J. B. (2003). Fucose: Biosynthesis and biological function in mammals. *Glycobiology, 13*, 41R–53R.
93. Keeley, T. S., Yang, S., & Lau, E. (2019). The diverse contributions of Fucose linkages in cancer. *Cancers, 11*, E1241.
94. Zhou, J., et al. (2021). CircularRNA circPARP4 promotes glioblastoma progression through sponging miR-125a-5p and regulating FUT4. *American Journal of Cancer Research, 11*, 138–156.
95. Read, T.-A., et al. (2009). Identification of CD15 as a marker for tumor-propagating cells in a mouse model of medulloblastoma. *Cancer Cell, 15*, 135–147.
96. Kim, K.-J., et al. (2011). The presence of stem cell marker-expressing cells is not prognostically significant in glioblastomas. *Neuropathology: Official Journa; of Japan Soceity of Neuropathology, 31*, 494–502.
97. Zhang, Y., Zhang, D., Lv, J., Wang, S., & Zhang, Q. (2018). MiR-125a-5p suppresses bladder cancer progression through targeting FUT4. *Biomedicine & Pharmacotherapy, 108*, 1039–1047.
98. Li, Y., et al. (2017). Tumor-suppressive miR-26a and miR-26b inhibit cell aggressiveness by regulating FUT4 in colorectal cancer. *Cell Death & Disease, 8*, e2892.
99. Cheng, L., et al. (2013). FUT family mediates the multidrug resistance of human hepatocellular carcinoma via the PI3K/Akt signaling pathway. *Cell Death & Disease, 4*, e923.
100. Yang, X., Liu, S., & Yan, Q. (2013). Role of fucosyltransferase IV in epithelial-mesenchymal transition in breast cancer cells. *Cell Death & Disease, 4*, e735.
101. Liu, B., et al. (2019). MiR-29b/Sp1/FUT4 axis modulates the malignancy of leukemia stem cells by regulating fucosylation via Wnt/β-catenin pathway in acute myeloid leukemia. *Journal of Experimental & Clinical Cancer Research, 38*, 200.
102. Chen, R., Zhang, X., & Wang, C. (2020). LncRNA HOXB-AS1 promotes cell growth in multiple myeloma via FUT4 mRNA stability by ELAVL1. *Journal of Cellular Biochemistry, 121*, 4043–4051.
103. Chen, X., Li, L. Q., Qiu, X., & Wu, H. (2019). Long non-coding RNA HOXB-AS1 promotes proliferation, migration and invasion of glioblastoma cells via HOXB-AS1/miR-885-3p/HOXB2 axis. *Neoplasma, 66*, 386–396.
104. Desiderio, V., et al. (2015). Increased fucosylation has a pivotal role in invasive and metastatic properties of head and neck cancer stem cells. *Oncotarget, 6*, 71–84.
105. Jassam, S. A., et al. (2017). CD15s/CD62E interaction mediates the adhesion of non-small cell lung cancer cells on brain endothelial cells: Implications for cerebral metastasis. *International Journal of Molecular Sciences, 18*, E1474.
106. Jassam, S. A., Maherally, Z., Ashkan, K., Pilkington, G. J., & Fillmore, H. L. (2019). Fucosyltransferase 4 and 7 mediates adhesion of non-small cell lung cancer cells to brain-derived endothelial cells and results in modification of the blood-brain-barrier: In vitro investigation of CD15 and CD15s in lung-to-brain metastasis. *Journal of Neuro-Oncology, 143*, 405–415.
107. Carrascal, M. A., et al. (2018). Inhibition of fucosylation in human invasive ductal carcinoma reduces E-selectin ligand expression, cell proliferation, and ERK1/2 and p38 MAPK activation. *Molecular Oncology, 12*, 579–593.
108. Ruan, W., et al. (2021). FUT11 is a target gene of HIF1α that promotes the progression of hepatocellular carcinoma. *Cell Biology International, 45*, 2275–2286.
109. Cao, W., et al. (2021). Hypoxia-related gene FUT11 promotes pancreatic cancer progression by maintaining the stability of PDK1. *Frontiers in Oncology, 11*, 675991.
110. Zodro, E., et al. (2013). FUT11 as a potential biomarker of clear cell renal cell carcinoma progression based on meta-analysis of gene expression data. *Tumour Biology, 35*, 2607–2617.

111. Shergalis, A., Bankhead, A., Luesakul, U., Muangsin, N., & Neamati, N. (2018). Current challenges and opportunities in treating glioblastoma. *Pharmacological Reviews, 70*, 412–445.
112. Taniguchi, N., & Kizuka, Y. (2015). Glycans and cancer: Role of N-glycans in cancer biomarker, progression and metastasis, and therapeutics. *Advances in Cancer Research, 126*, 11–51.
113. Castriconi, R., et al. (2004). Identification of 4Ig-B7-H3 as a neuroblastoma-associated molecule that exerts a protective role from an NK cell-mediated lysis. *Proceedings of the National Academy of Sciences of the United States of America, 101*, 12640–12645.
114. Lemke, D., et al. (2012). Costimulatory protein 4IgB7H3 drives the malignant phenotype of glioblastoma by mediating immune escape and invasiveness. *Clinical Cancer Research : An Official Journal of the American Association for Cancer Research, 18*, 105–117.
115. Sun, Z., et al. (2019). Mucin O-glycosylating enzyme GALNT2 facilitates the malignant character of glioma by activating the EGFR/PI3K/Akt/mTOR axis. *Clinical Science London England, 1979*(133), 1167–1184.
116. Hussain, M. R. M., Hoessli, D. C., & Fang, M. (2016). N-acetylgalactosaminyltransferases in cancer. *Oncotarget, 7*, 54067–54081.
117. Banford, S., & Timson, D. J. (2017). UDP-N-acetyl-D-galactosamine:Polypeptide N-acetylgalactosaminyltransferase- 6 (pp-GalNAc-T6): Role in cancer and prospects as a drug target. *Current Cancer Drug Targets, 17*, 53–61.
118. Lin, W.-R., & Yeh, C.-T. (2020). GALNT14: An emerging marker capable of predicting therapeutic outcomes in multiple cancers. *International Journal of Molecular Sciences, 21*, E1491.
119. Calvo, P., Reglero, A., & Cabezas, J. A. (1978). Purification and properties of beta-N-acetylhexosaminidase from the mollusc Helicella ericetorum Müller. *The Biochemical Journal, 175*, 743–750.
120. Mahuran, D. J. (1995). Beta-hexosaminidase: Biosynthesis and processing of the normal enzyme, and identification of mutations causing Jewish Tay-Sachs disease. *Clinical Biochemistry, 28*, 101–106.
121. Tifft, C. J., & Proia, R. L. (1997). The beta-hexosaminidase deficiency disorders: Development of a clinical paradigm in the mouse. *Annals of Medicine, 29*, 557–561.
122. Mian, N., & Cowen, D. M. (1974). Glycosidases in normal and dimethylhydrazine-treated rats and mice with special reference to the colonic tumours. *British Journal of Cancer, 29*, 438–446.
123. Bosmann, H. B., & Hall, T. C. (1974). Enzyme activity in invasive tumors of human breast and colon. *Proceedings of the National Academy of Sciences of the United States of America, 71*, 1833–1837.
124. Gil-Martín, E., Rodríguez-Berrocal, F. J., Páez de la Cadena, M., & Fernández-Briera, A. (1999). N-acetyl-beta-hexosaminidase activity and isoenzymes in human gastric adenocarcinoma. *Oncology, 56*, 142–154.
125. Chatterjee, S. K., Chowdhury, K., Bhattacharya, M., & Barlow, J. J. (1982). Beta-hexosaminidase activities and isoenzymes in normal human ovary and ovarian adenocarcinoma. *Cancer, 49*, 128–135.
126. Kawahara, R., et al. (2021). The complexity and dynamics of the tissue Glycoproteome associated with prostate cancer progression. *Molecular & Cellular Proteomics, 20*, 100026.
127. Kawahara, R., et al. (2018). Distinct urinary glycoprotein signatures in prostate cancer patients. *Oncotarget, 9*, 33077–33097.
128. Costanzi, E., et al. (2014). Hypermethylation contributes to down-regulation of lysosomal β-hexosaminidase α subunit in prostate cancer cells. *Biochimie, 101*, 75–82.
129. Alhadeff, J. A., Prorok, J. J., Dura, P. A., Plucinsky, M. C., & Beesburg, R. Y. (1984). Atypical beta-hexosaminidase in sera of cancer patients with liver metastases. *Cancer Research, 44*, 5422–5426.
130. Thaysen-Andersen, M., et al. (2015). Human neutrophils secrete bioactive paucimannosidic proteins from azurophilic granules into pathogen-infected sputum. *The Journal of Biological Chemistry, 290*, 8789–8802.

131. Schnaar, R. L., Gerardy-Schahn, R., & Hildebrandt, H. (2014). Sialic acids in the brain: Gangliosides and polysialic acid in nervous system development, stability, disease, and regeneration. *Physiological Reviews, 94*, 461–518.
132. Tjondro, H. C., Loke, I., Chatterjee, S., & Thaysen-Andersen, M. (2019). Human protein paucimannosylation: Cues from the eukaryotic kingdoms. *Biological Reviews of the Cambridge Philosophical Society, 94*, 2068–2100.
133. Everest-Dass, A. V., et al. (2012). Comparative structural analysis of the glycosylation of salivary and buccal cell proteins: Innate protection against infection by Candida albicans. *Glycobiology, 22*, 1465–1479.
134. Loke, I., Østergaard, O., Heegaard, N. H. H., Packer, N. H., & Thaysen-Andersen, M. (2017). Paucimannose-rich N-glycosylation of spatiotemporally regulated human neutrophil elastase modulates its immune functions. *Molecular & Cellular Proteomics, 16*, 1507–1527.
135. Olczak, M., & Watorek, W. (2002). Structural analysis of N-glycans from human neutrophil azurocidin. *Biochemical and Biophysical Research Communications, 293*, 213–219.
136. Howard, D. R., Natowicz, M., & Baenziger, J. U. (1982). Structural studies of the endoglycosidase H-resistant oligosaccharides present on human beta-glucuronidase. *The Journal of Biological Chemistry, 257*, 10861–10868.
137. Ravnsborg, T., Houen, G., & Højrup, P. (2010). The glycosylation of myeloperoxidase. *Biochimica et Biophysica Acta, 1804*, 2046–2053.
138. Loke, I., Packer, N. H., & Thaysen-Andersen, M. (2015). Complementary LC-MS/MS-based N-glycan, N-Glycopeptide, and intact N-glycoprotein profiling reveals unconventional Asn71-glycosylation of human neutrophil Cathepsin G. *Biomolecules, 5*, 1832–1854.
139. Balog, C. I. A., et al. (2012). N-glycosylation of colorectal cancer tissues: A liquid chromatography and mass spectrometry-based investigation. *Molecular & Cellular Proteomics, 11*, 571–585.
140. Chatterjee, S., et al. (2019). Protein Paucimannosylation is an enriched N-glycosylation signature of human cancers. *Proteomics, 19*, e1900010.
141. Zipser, B., Bello-DeOcampo, D., Diestel, S., Tai, M.-H., & Schmitz, B. (2012). Mannitou monoclonal antibody uniquely recognizes Paucimannose, a marker for human cancer, Stemness, and inflammation. *Journal of Carbohydrate Chemistry, 31*, 504–518.
142. Dahmen, A.-C., et al. (2015). Paucimannosidic glycoepitopes are functionally involved in proliferation of neural progenitor cells in the subventricular zone. *Glycobiology, 25*, 869–880.
143. Tropak, M. B., Blanchard, J. E., Withers, S. G., Brown, E. D., & Mahuran, D. (2007). High-throughput screening for human lysosomal beta-N-acetyl hexosaminidase inhibitors acting as pharmacological chaperones. *Chemistry & Biology, 14*, 153–164.
144. Braña, M. F., et al. (1993). Bis-naphthalimides: A new class of antitumor agents. *Anti-Cancer Drug Design, 8*, 257–268.
145. Brana, M. F., & Ramos, A. (2001). Naphthalimides as anticancer agents: Synthesis and biological activity. *Current Medicinal Chemistry - Anti-Cancer Agents, 1*, 237–255.
146. Awada, A., et al. (2003). An EORTC-ECSG phase I study of LU 79553 administered every 21 or 42 days in patients with solid tumours. *European Journal of Cancer Oxford England, 1990*(39), 742–747.
147. Matsuzaki, H., et al. (2016). Novel Hexosaminidase-targeting fluorescence probe for visualizing human colorectal cancer. *Bioconjugate Chemistry, 27*, 973–981.
148. Fiaux, H., et al. (2008). Functionalized pyrrolidine inhibitors of human type II alpha-mannosidases as anti-cancer agents: Optimizing the fit to the active site. *Bioorganic & Medicinal Chemistry, 16*, 7337–7346.
149. Hamester, F., et al. (2019). Prognostic relevance of the Golgi mannosidase MAN1A1 in ovarian cancer: Impact of N-glycosylation on tumour cell aggregation. *British Journal of Cancer, 121*, 944–953.
150. Li, M., et al. (2020). Characterization of hepatitis B virus infection and viral DNA integration in non-Hodgkin lymphoma. *International Journal of Cancer, 147*, 2199–2209.

151. Zhang, S., Zhang, Q., Yin, J., & Wu, X. (2019). Overlapped differentially expressed genes between acute lymphoblastic leukemia and chronic lymphocytic leukemia revealed potential key genes and pathways involved in leukemia. *Journal of Cellular Biochemistry, 120*, 15980–15988.
152. Wielgat, P., et al. (2006). Activity of lysosomal exoglycosidases in human gliomas. *Journal of Neuro-Oncology, 80*, 243–249.
153. Urbanelli, L., et al. (2011). Human lysosomal α-D-mannosidase regulation in promyelocytic leukaemia cells. *Bioscience Reports, 31*, 477–487.
154. Boscher, C., Dennis, J. W., & Nabi, I. R. (2011). Glycosylation, galectins and cellular signaling. *Current Opinion in Cell Biology, 23*, 383–392.
155. Oh, S., Kim, H., Nam, K., & Shin, I. (2017). Glutl promotes cell proliferation, migration and invasion by regulating epidermal growth factor receptor and integrin signaling in triple-negative breast cancer cells. *BMB Reports, 50*, 132–137.
156. Kizuka, Y., & Taniguchi, N. (2016). Enzymes for N-glycan branching and their genetic and nongenetic regulation in cancer. *Biomolecules, 6*, E25.
157. Takamatsu, S., et al. (1999). Unusually high expression of N-acetylglucosaminyltransferase-IVa in human choriocarcinoma cell lines: A possible enzymatic basis of the formation of abnormal biantennary sugar chain. *Cancer Research, 59*, 3949–3953.
158. Guo, Y., et al. (2014). Let-7c inhibits metastatic ability of mouse hepatocarcinoma cells via targeting mannoside acetylglucosaminyltransferase 4 isoenzyme a. *The International Journal of Biochemistry & Cell Biology, 53*, 1–8.
159. Kizuka, Y., et al. (2014). Epigenetic regulation of a brain-specific glycosyltransferase N-acetylglucosaminyltransferase-IX (GnT-IX) by specific chromatin modifiers. *The Journal of Biological Chemistry, 289*, 11253–11261.
160. Lange, T., et al. (2012). Human prostate cancer in a clinically relevant xenograft mouse model: Identification of β(1,6)-branched oligosaccharides as a marker of tumor progression. *Clinical Cancer Research : An Official Journal of the American Association for Cancer Research, 18*, 1364–1373.
161. Marhuenda, E., et al. (2021). Glioma stem cells invasive phenotype at optimal stiffness is driven by MGAT5 dependent mechanosensing. *Journal of Experimental & Clinical Cancer Research, 40*, 139.
162. Granovsky, M., et al. (2000). Suppression of tumor growth and metastasis in Mgat5-deficient mice. *Nature Medicine, 6*, 306–312.
163. Lagana, A., et al. (2006). Galectin binding to Mgat5-modified N-glycans regulates fibronectin matrix remodeling in tumor cells. *Molecular and Cellular Biology, 26*, 3181–3193.
164. Pinho, S. S., et al. (2009). The role of N-acetylglucosaminyltransferase III and V in the post-transcriptional modifications of E-cadherin. *Human Molecular Genetics, 18*, 2599–2608.
165. Zhao, Y., et al. (2008). Branched N-glycans regulate the biological functions of integrins and cadherins. *The FEBS Journal, 275*, 1939–1948.
166. Pinho, S. S., et al. (2012). Loss and recovery of Mgat3 and GnT-III mediated E-cadherin N-glycosylation is a mechanism involved in epithelial-mesenchymal-epithelial transitions. *PLoS One, 7*, e33191.
167. Miyagi, T., & Yamaguchi, K. (2012). Mammalian sialidases: Physiological and pathological roles in cellular functions. *Glycobiology, 22*, 880–896.
168. Haxho, F., Neufeld, R. J., & Szewczuk, M. R. (2016). Neuraminidase-1: A novel therapeutic target in multistage tumorigenesis. *Oncotarget, 7*, 40860–40881.
169. Akasov, R., et al. (2016). Sialylation transmogrifies human breast and pancreatic cancer cells into 3D multicellular tumor spheroids using cyclic RGD-peptide induced self-assembly. *Oncotarget, 7*, 66119–66134.
170. Hou, G., et al. (2016). Neuraminidase 1 (NEU1) promotes proliferation and migration as a diagnostic and prognostic biomarker of hepatocellular carcinoma. *Oncotarget, 7*, 64957–64966.

171. Silvestri, I., et al. (2014). Sialidase NEU4 is involved in glioblastoma stem cell survival. *Cell Death & Disease, 5*, e1381.
172. Song, Z. (2013). Roles of the nucleotide sugar transporters (SLC35 family) in health and disease. *Molecular Aspects of Medicine, 34*, 590–600.
173. Hediger, M. A., Clémençon, B., Burrier, R. E., & Bruford, E. A. (2013). The ABCs of membrane transporters in health and disease (SLC series): Introduction. *Molecular Aspects of Medicine, 34*, 95–107.
174. Hadley, B., et al. (2019). Nucleotide sugar transporter SLC35 family structure and function. *Computational and Structural Biotechnology Journal, 17*, 1123–1134.
175. Handford, M., Rodriguez-Furlán, C., & Orellana, A. (2006). Nucleotide-sugar transporters: Structure, function and roles in vivo. *Brazilian Journal of Medical and Biological Research: Revista brasileira de pesquisas medicas e biologicas, 39*, 1149–1158.
176. Parker, J. L., Corey, R. A., Stansfeld, P. J., & Newstead, S. (2019). Structural basis for substrate specificity and regulation of nucleotide sugar transporters in the lipid bilayer. *Nature Communications, 10*, 4657.
177. Sosicka, P., Jakimowicz, P., Olczak, T., & Olczak, M. (2014). Short N-terminal region of UDP-galactose transporter (SLC35A2) is crucial for galactosylation of N-glycans. *Biochemical and Biophysical Research Communications, 454*, 486–492.
178. Büll, C., Stoel, M. A., den Brok, M. H., & Adema, G. J. (2014). Sialic acids sweeten a tumor's life. *Cancer Research, 74*, 3199–3204.
179. Rodrigues, E., & Macauley, M. S. (2018). Hypersialylation in cancer: Modulation of inflammation and therapeutic opportunities. *Cancers, 10*, E207.
180. Kotolloshi, R., Hölzer, M., Gajda, M., Grimm, M.-O. & Steinbach, D. SLC35F2, a transporter sporadically mutated in the untranslated region, promotes growth, migration, and invasion of bladder cancer cells. *Cells* 10, E80 (2021).
181. Bu, L., Jiang, G., Yang, F., Liu, J., & Wang, J. (2011). Highly expressed SLC35F2 in non-small cell lung cancer is associated with pathological staging. *Molecular Medicine Reports, 4*, 1289–1293.
182. Huang, E. Y., et al. (2018). A common regulatory variant in SLC35B4 influences the recurrence and survival of prostate cancer. *Journal of Cellular and Molecular Medicine, 22*, 3661–3670.
183. Harduin-Lepers, A., et al. (2001). The human sialyltransferase family. *Biochimie, 83*, 727–737.
184. Paulson, J. C., & Rademacher, C. (2009). Glycan terminator. *Nature Structural & Molecular Biology, 16*, 1121–1122.
185. Li, Y., & Chen, X. (2012). Sialic acid metabolism and sialyltransferases: Natural functions and applications. *Applied Microbiology and Biotechnology, 94*, 887–905.
186. Harduin-Lepers, A., et al. (2008). Evolutionary history of the alpha2,8-sialyltransferase (ST8Sia) gene family: Tandem duplications in early deuterostomes explain most of the diversity found in the vertebrate ST8Sia genes. *BMC Evolutionary Biology, 8*, 258.
187. Dobie, C., & Skropeta, D. (2021). Insights into the role of sialylation in cancer progression and metastasis. *British Journal of Cancer, 124*, 76–90.
188. Milcheva, R., Georgieva, A. K., Todorova, K. S. & Petkova, S. L. (2022). *Expression of sialyltransferases from the ST3Gal, ST6Gal and ST6GalNAc families in mouse skeletal muscle and mouse C2C12 myotube cell cultures*. https://doi.org/10.21203/rs.3.rs-316783/v1.
189. Audry, M., et al. (2011). Current trends in the structure-activity relationships of sialyltransferases. *Glycobiology, 21*, 716–726.
190. Bordron, A., et al. (2018). Resistance to complement activation, cell membrane hypersialylation and relapses in chronic lymphocytic leukemia patients treated with rituximab and chemotherapy. *Oncotarget, 9*, 31590–31605.
191. Pietrobono, S., et al. (2020). ST3GAL1 is a target of the SOX2-GLI1 transcriptional complex and promotes melanoma metastasis through AXL. *Nature Communications, 11*, 5865.

192. Lin, W.-D., et al. (2021). Sialylation of CD55 by ST3GAL1 facilitates immune evasion in cancer. *Cancer Immunology Research, 9*, 113–122.
193. Ma, X., et al. (2016). Functional roles of sialylation in breast cancer progression through miR-26a/26b targeting ST8SIA4. *Cell Death & Disease, 7*, e2561.
194. Guerrero, P. E., et al. (2020). Knockdown of α2,3-Sialyltransferases impairs pancreatic cancer cell migration, invasion and E-selectin-dependent adhesion. *International Journal of Molecular Sciences, 21*, E6239.
195. Vojta, A., Samaržija, I., Bočkor, L., & Zoldoš, V. (2016). Glyco-genes change expression in cancer through aberrant methylation. *Biochimica et Biophysica Acta, 1860*, 1776–1785.
196. Horvat, T., Mužinić, A., Barišić, D., Bosnar, M. H., & Zoldoš, V. (2012). Epigenetic modulation of the HeLa cell membrane N-glycome. *Biochimica et Biophysica Acta, 1820*, 1412–1419.
197. Perez, M., Chakraborty, A., Lau, L. S., Mohammed, N. B. B., & Dimitroff, C. J. (2021). Melanoma-associated glycosyltransferase GCNT2 as an emerging biomarker and therapeutic target. *The British Journal of Dermatology, 185*, 294–301.
198. Klasić, M., et al. (2016). DNA hypomethylation upregulates expression of the MGAT3 gene in HepG2 cells and leads to changes in N-glycosylation of secreted glycoproteins. *Scientific Reports, 6*, 24363.

Chapter 5
Post-translational Modification in Muscular Dystrophies

Martina Sandonà and Valentina Saccone

Abstract Muscular dystrophies are a complex group of inherited neuromuscular disorders that progressively lead to a loss of muscle fibers and mobility and muscle weakness; over time, they evolve to an increasing level of disability. Muscular dystrophies are mostly caused by genetic mutations in proteins responsible for maintaining sarcolemma structures, such as an absence or reductions of dystrophin expression, conditions which are strictly related to muscular disorders that affect most people with this disease. Along the years, with the recent advances in the understanding of muscular dystrophies, it has been shown that many changes in Post-Translational Modifications (PTMs) of muscle proteins are associated with muscular dystrophies, wherein pathogenic alterations in the modulation of these muscle proteins are directly related to the incidence of this disease. An increase in the identification of the genetic bases and molecular mechanisms involved in the most common form of muscular dystrophies, including PTMs changes, holds potential to develop new therapeutic approaches. In this chapter we will describe the most common muscular dystrophies and changes in PTM processes such as phosphorylation and glycosylation that are very important in the evolution of the disease, highlighting the lack of mass spectrometry-based (MS-based) studies of these PTMs, suggesting that the application of this technique could reveal important informations about the molecular mechanisms of muscular dystrophies.

Keywords Muscular dystrophies · Post-translational modifications · Duchenne muscular dystrophy · Mass spectrometry

M. Sandonà
Istituto di Ricovero e Cura a Carattere Scientifico (IRCCS), Fondazione Santa Lucia, Rome, Italy

V. Saccone (✉)
Istituto di Ricovero e Cura a Carattere Scientifico (IRCCS), Fondazione Santa Lucia, Rome, Italy

Dipartimento di Scienze della Vita e della Sanità Pubblica, Università Cattolica del Sacro Cuore, Rome, Italy

V. Corasolla Carregari (ed.), *Understanding PTMs in Neurodegenerative Diseases*, Advances in Experimental Medicine and Biology 1382,
https://doi.org/10.1007/978-3-031-05460-0_5

5.1 Introduction

Muscular dystrophies (MDs) are a class of complex, inherited neuromuscular disorders characterized by progressive muscle wasting and weakness of varying distribution and severity. They can be subdivided into several groups, including congenital forms. The main cause of MDs is due to genetic mutations in proteins responsible for maintaining the sarcolemma structure, as in the case of the two most severe MDs: Duchenne (DMD) and Becker (BMD). Both are X-linked, recessive disorders caused by mutations in the dystrophin gene, resulting in reduced or abolished expression of dystrophin or an expression of truncated forms of the dystrophin protein. More specifically, the absence of dystrophin causes DMD, whereas mutations causing reduced expression or truncation of the dystrophin protein cause BMD. Both DMD and BMD are characterized by a progressive skeletal muscle degeneration, weakness, loss of ambulation, and early death due to cardiorespiratory insufficiency and a broad series of secondary effects [1]. Other inherited muscular dystrophies include: Emery-Dreifuss, facioscapulohumeral, oculopharyngeal, congenital forms, and limb-girdle, which are caused by mutations in genes that encode sarcolemma, extracellular matrix, sarcomeric, and nuclear envelope proteins [2–5].

Identifications of the genetic bases of the most common forms of muscular dystrophy and improved understanding of the mechanisms that underlie muscular dystrophies have already resulted in the development of new therapeutic approaches, several of which are already in clinical trials. Some of these clinical trials target particular mutations, whereas others focus on secondary aspects of muscle degeneration.

Most recent advances in muscular disorders have also focused on post-translational processing of muscle proteins, which have been shown to modulate muscle functions and the incidence of muscle diseases [6]. These post-translational processes refer to the covalent and enzymatic modification of proteins following protein biosynthesis, having been only recently shown to be primarily involved in the pathogenesis of muscular dystrophies. Post-Translational Modifications (PTMs) include enzymatic cleavage of peptide bonds and the covalent additions of chemical groups, lipids, carbohydrates, or even entire proteins to amino acid side chains. These chemical modifications change protein structures influencing their biological activity [7].

5.2 PTMs and Duchenne Muscular Dystrophy

Understanding the impact of dystrophin deficiency in a murine model of DMD–termed mdx mice–is made much more difficult due to the additional loss of the entire protein complex that interacts with dystrophin on the surface of muscle cells [8]. In particular, Brenman et al. demonstrated that neuronal nitric oxide synthase (nNOS) interacts with the dystrophin complex on the muscle cell membrane, and that the absence of dystrophin in muscles leads to a loss of nNOS protein expression and activity [9]. A reduction in nNOS expression leads to a decrease in nitric oxide (NO) bioavailability in skeletal muscles which can be associated with chronic inflammation, tissue degeneration, and deficiencies in regeneration [10].

NO is an important signaling molecule that contributes to the production of cyclic guanosine monophosphate (cGMP) and the S-nitrosylation of thiol residues of cysteine groups, as well as in histone deacetylase enzymes as HDAC2 [11]. This is shown in mdx mice, as the deregulation of nNOS leads to low levels of S-nitrosylated cysteines of HDAC2, as observed by immunoprecipitation with an anti-S-nitroso-cysteine-specific antibody; these lower S-nitrosylation levels in turn increase HDAC2 activity, which is known to be responsible for the deregulated gene expression in dystrophic muscles [12].

Moreover, different evidences highlighted the role of NO as an epigenetic modulator of chromatin in DMD muscle; loss of nNOS from dystrophic muscle was associated with increased levels of histone acetylation [13]. Histone H3 acetylation at lysine 9 was especially elevated in the mdx mouse model compared to healthy control muscles. Those levels were normalized in dystrophic muscles by transfection with endothelial NOS, supporting the idea that the loss of nNOS contributed to the histone modification changes.

Detyrosination of alpha-tubulin is a post-translational modification characterized by the cleavage of a C-terminal tyrosine residue by a carboxypeptidase (TCP) [14, 15], influencing X-ROS signaling, contraction speed, and cytoskeletal mechanics, in addition to regulating cytosolic calcium concentrations [16]. Considering its central role in muscle organization and in muscle damage, different studies reported an increase of detyrosinated alpha-tubulin in dystrophic skeletal [17, 18] and cardiac [19] muscles. Moreover, a more recent study using MALDI-TOF mass spectrometry, revealed that detyrosinated alpha-tubulin in BMD muscles remained unchanged, defining the dysregulation of this post-translational modification as one of the central nodes of DMD/BMD phenotypes [20].

PTMs of alpha-tubulin have also been shown to stabilize microtubule (MT) filaments by limiting disassembly [21] and can also increase their association with other cytoskeletal elements [22]. MT detyrosination affects both cytoskeletal stiffness and the contraction velocity of muscles, and is therefore involved in mechanotransduction [16]. Given its ability to alter mechanical properties of the cytoskeleton, detyrosination could become a novel target for therapeutic strategies in DMD [16]. Moreover, elevated MT density and detyrosination have been described in other muscular dystrophies [23, 24] and cardiac diseases [19, 25], and therefore its manipulation appears useful also for the treatment of other muscular diseases.

DMD is characterized by serious cardiac complications, which present themselves at advanced stages of the disease. In mdx mice, cardiomyopathy becomes prominent at 8 months of age, though the first signs are observed at 3 months of age. While cardiomyopathy appears in adulthood in mdx mice, a lack of dystrophin is present throughout the lifetime of these animals; therefore, the absence of this protein influences the hearts of young mdx mice, even if they do not yet develop cardiomyopathy. Jung et al. linked dystrophin deficiency in hearts to altered Ca^{2+}

homeostasis, which is caused by the activation of sarcoplasmic reticulum (SR) Ca^{2+} release [26, 27]. Using young cardiomyocytes from mdx mice, oxidative stress was demonstrated to be one of the initial problems in dystrophic myocytes. ROS production is abnormally high in mdx mice, and this activates CamKII in a Ca^{2+} independent way [28]. This event leads to a CaMKII-dependent phosphorylation of ryanodine receptors (RyR), with its subsequent sensitization. This mechanism could explain the susceptibility of dystrophic cardiomyocytes to stress at early stages of the disease [29].

During the progression of DMD, but before the heart failure, RyR undergoes phosphorylation by CaMKII and PKA and undergoes nitrosation by eNOS, nNOS, and iNOS; these PTMs increase RyR sensitivity to Ca^{2+}, contributing to cardiac deterioration [29–31]. Moreover, at terminal stages of DMD, several post-translational modifications further increase RyR sensitivity, leading to an increase of SR Ca^{2+} leakage which in turn contributes to an overload of cytosolic and mitochondrial Ca^{2+} and subsequent activation of necrotic and apoptotic pathways and a loss of myocyte functions [29].

In addition, Bellinger et al. [32], demonstrated that in skeletal muscle of dystrophic mice, RyR-1 presented structural and functional defects, which contributed to altered Ca^{2+} homeostasis. In particular, they observed increased S-nitrosylation of RyR-1, which correlated with the progression of the disease and with the alteration of NO signaling in skeletal muscle. The PTMs of RyR lead to a depletion of the channel complex of FKBP12 (also known as calstabin-1), resulting in 'leaky' channels. This contributes to a lowered accessibility of Ca^{2+} for contraction, and therefore results in muscle weakness [32].

5.3 Glycosylation of α-Dystroglycan in MDs

More than half of proteins are known to undergo glycosylation, a post-translational modification that affects protein stability and conformation, additionally playing roles in molecular recognition events that occur in numerous processes characterized by intercellular communication such as: infection, inflammation and metastasis, differentiation, development, and many others [33].

The dystrophin-associate protein complex (DAPC) is present in muscle fibers and provides mechanical support during contraction/relaxation and is understood to function in signal transduction [34]. The DAPC is composed of dystrophin, cytoskeletal proteins, and proteins that themselves form 3 separate complexes: the dystroglycan subcomplex, the sarcoglycan-sarcospan subcomplex, and the cytoplasmic subcomplex [35, 36]. The DAPC includes dystroglycans, sarcoglycans, sarcospan, syntrophins, and dystrobrevin; and mutations in any of these proteins may result in loss of membrane integrity [37, 38]. Moreover, abnormalities in other glycosyltransferases that support O-mannosylation of α-dystroglycan cause other, non-DMD/BMD forms of muscular dystrophy [34, 39].

Dystroglycan (DG) is a cell surface receptor for several extracellular matrix proteins. DG is the central transmembrane protein in the dystrophin–glycoprotein complex (DGC), linking dystrophin which is expressed in the intracellular cytoskeleton to proteins of the extracellular matrix [40]. Glycosylation of DG is necessary for its ligand binding activity, and hypoglycosylation is associated with a group of muscular dystrophies that often involve abnormalities [40–42]. Dystroglycan is synthesized as a molecular precursor that is post-translationally cleaved into two subunits. The cleavage of DC generates the transmembrane protein α-dystroglycan (α-DG), which interacts directly with dystrophin at its intracellular domain; while outside the plasmatic membrane, it binds to extracellular α-DG. In turn, α-DG binds to the extracellular matrix (ECM) protein laminin as well as agrin, perlecan, neurexin, and pikachurin [40, 43, 44]. Both α- and β-dystroglycans undergo extensive glycosylation, with their modification varying by tissue; in particular, α-dystroglycan is present in several tissues, including skeletal muscle, heart, nerve and brain [40, 45, 46].

The glycosylation of α-DG occurs across three different domains: a central, heavily O-glycosylated mucin domain and two globular, N-glycosylated domains at the N- and C-terminal [47, 48]. These PTMs of α-DG alter its structure, localization, and binding; therefore, the tissue-specific glycosylated forms of α-DG reflect its different functions in each tissue and at each stage of development. For example, the O-mannosylation of α-DG is required for laminin binding of the protein in different muscular diseases [49–52].

In dystrophic hearts, the different forms of glycosylated α-DG do not influence its laminin binding activity, but rather modify its localization at the sarcolemma, contributing to the severity of the disease [53]. In skeletal muscle disease, the modified sialic acid of α-DG correlates with disease severity [54]; in a murine model, overexpression of Galgt2 (GalNAc transferase), an enzyme responsible for the glycosylation of glycoproteins, protected dystrophic skeletal muscle from degeneration [55]. Taken together, this evidence define the association of PTMs of α-DG with skeletal and cardiac muscle phenotypes in dystrophin-deficient mice (Table 5.1).

Table 5.1 List of the most important PTMs in Duchenne Muscular Dystrophy

PTM	Effects
S-nitrosylation	HDAC deregulation
Detyrosination of α-tubulin	Deregulation of: X-ROS signaling Contraction speed Cytoskeletal mechanisms Cytosolic Ca^{2+}
Phosphorylation, oxidation, and nitrosation of RyR	Increase of SR Ca^{2+} leak
α-DG glycosylation	Disease severity

5.4 PTMs and Other Muscular Dystrophies

Dystroglycanopathies can be given two classifications: diseases in which α-DG itself is mutated, and diseases in which proteins directly or indirectly involved in α-DG modification are mutated. Several forms of congenital muscular dystrophy are characterized by hypoglycosylation of α-dystroglycan and therefore they are described as congenital disorders of glycosylation. The lack of α-DG glycosylation inhibits its binding of extracellular matrix ligands; indeed, in all these dystroglycanopathies, the expression of the laminin α2 chain is reduced and the resulting skeletal muscle defects are similar to those observed in diseases in which laminin α2 chain is absent [51, 56]. To date, several glycosyltransferase genes have been involved in dystroglycanopathies:

- Protein O-mannosyl-transferase 1 (POMT1) and 2 (POMT2) are O-mannosyltransferases, whose dysregulation can cause a severe form of congenital muscular dystrophy named **Walker-Warburg Syndrome (WWS)** [57, 58]. These enzymes regulate the first step of O-mannosyl glycan synthesis, as proved by radioactive mannose incorporation into α-DG [59]. Indeed, WWS patients are characterized by a marked reduction in the glycosylated form of α-DG, particularly in peripheral nerves [60].
- Defects in protein O-mannose beta-1,2-N-acetylglucosaminyltransferase 1 (POMGnT1) cause **Muscle-Eye-Brain**(MEB) disease, a form of congenital muscular dystrophy characterized by brain malformations, ocular abnormalities, and muscle weakness [49]. POMGnT1 catalyzes the transfer of N-acetylglucosamine to O-linked mannose [61], and therefore mutations in the gene for this enzyme affect the glycosylation of α-DG in MEB [62].

As both POMT1 and POMGnT1 have O-mannosyl transferase activity, WWS and MEB therefore present similarities in their pathology, also supporting the fundamental role of glycosylated α-DG in muscular dystrophy.

- Fukutin gene mutations lead to **Fukuyama-type congenital muscular dystrophy**, characterized by severe muscle weakness, cardiomyopathy, and central nervous system defects [63]. Fukutin has homology with the fukutin family enzymes that modify glycolipids and glycoproteins; in this disease, the same absence of the glycosylated form of α-DG that is observed in MEB disease also occurs [64]. Glycoprotein enrichment and Western blot analyses suggest that fukutin may interact with the O-mannosyltransferases POMT1 and POMGnT1 to regulate α-DG glycosylation [51]. A homologue of fukutin (fukutin-related protein; FKRP) has been found mutated in children affected by congenital muscular dystrophy type 1C (**MDC1C**) and a less severe variant of MDC1C named limb-grid muscular dystrophy 2I (**LGMD2I**) [65, 66]. Western blot analyses of samples from MDC1C and LGMD2I patients show a loss of the higher molecular weight forms of α-dystroglycan; but, unlike the complete absence shown in MEB and FCMD, they present only reduced expression of α-dystroglycan [33].

- The LARGE gene encodes two glycosyltransferases, α3-xylosyltransferase and β3-glucoronyltransferase, and its mutation is responsible for **myodystrophy** (myd) [67, 68]. This pathology is characterized by defects in skeletal and cardiac muscles, eyes, and the central nervous system.
- In LARGE^{myd} mice, dystroglycan is incompletely glycosylated and thus cannot bind its extracellular ligands; in particular, a reduction of laminin binding occurs, causing a muscular dystrophy that is usually lethal in early adulthood [69, 70]. A mutation in LARGE also occurs in congenital muscular dystrophy type 1D (MDC1D), though the reduction of glycosylated α-DG is not as drastic [68]. In regards to LARGE overexpression, Campbell et al. demonstrated that α-dystroglycan becomes pathologically modified with the synthesis of species recognized by the IIH6 antigen [71, 72]. The LARGE-induced hyperglycosylation of α-DG is followed by increased laminin binding.
- ISPD encodes isoprenoid synthase domain-containing protein, and its mutations were described as another factor involved in WWS and other dystroglycanopathies [52, 73, 74]. ISPD is involved in the synthesis of tandem Rbo5P in the sugar chain of α-DG, which is indispensable for the functional maturation of α-DG [75]. In particular, Kanagawa et al. made use of MALDI-TOF and nanoelectrospray ionization (nanoESI) mass-spectrometry to study PTMs of proteins, and demonstrated that ISPD as a CDP-Rbo synthase uses Rbo5P and CTP, while fukutin and fukutin-related protein act as Rbo5P transferases, creating the tandem repeats of Rbo5P on α-DG that are necessary for the formation of the ligand-binding moiety [75]. Three α-dystroglycanopathy-causing proteins, namely ISPD, fukutin, and FKRP, seem to be involved in Rbo5P glycosylation (Fig. 5.1).

5.5 Muscular Dystrophy Treatments and PTMs

As previously mentioned, muscular dystrophies are heterogeneous and complex diseases, caused by genetic defects and often associated with a deregulation of post-translational modifications. For this reason, several studies have focused on the development of therapeutic strategies based not only on restoring genetic defects but also deregulated PTMs to ameliorate muscle pathology.

For example, restoring the activity of the glycosylating enzyme CT-GalNAc/Galgt2 (cytotoxic T cell β1,4-*N*-acetylgalactosamine transferase) represents one of the strategies to recover dystrophic muscle. In mdx and other muscular diseases, ectopic expression of the normally synaptic CT-GalNAc transferase leads to a recovery of utrophin, sarcoglycan, dystrobrevin, and glycosylated dystroglycan expression [76, 77]. In addition, Martin et al. [55] demonstrated that the overexpression of Galgt2 in mdx as well as wt mice protects muscles from damage, preventing the loss of force during eccentric contractions [55].

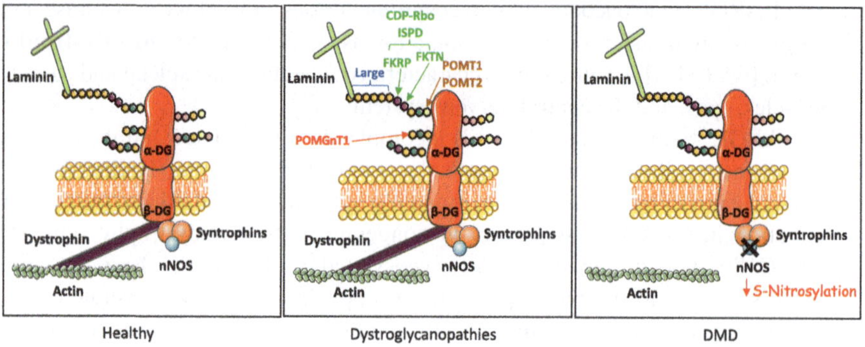

Fig. 5.1 Representation of molecular organization of the dystrophin–glycoprotein complex (DGC) in sarcolemma under physiological and PTM-pathological conditions. Representations are of (left) healthy skeletal muscle tissue, (middle) disorders associated with PTMs in α-DG called dystroglycanopathies, and a representation of DMD skeletal muscle tissue (right). In particular, α-DG glycosylation and the relative enzymes involved in these PTMs in α-DG are shown. Mutations in these enzymes lead to severe diseases such as Fukuyama-type congenital muscular dystrophy, MDC1C, LGMD2I, Walker-Warburg Syndrome and Muscle-eye-brain disease as described in the text. A lack of dystrophin in muscle tissue leads to non-functional nNOS, thereby decreasing HDAC S-Nitrosylation

The glycosylated form of α-DG was also able to be restored by the addition of cytidine diphosphate-ribitol (CDP-Rbo). Since fukutin and fukutin-related protein sequentially act on Rbo5P transferases that use CDP-Rbo to form a tandem structure on α-DG that is necessary for its ligand activity, Kanagawa et al. added CDP-Rbo to ISPD-deficient cell culture medium and saw a recovery of glycosylation of α-DG underling the pivotal role of CDP-Rbo in this system and in dystroglycanopathies [75].

Another PTM that could be targeted for therapeutic strategies in dystrophic muscle is the detyrosination of alpha-tubulin; indeed, the reduction of detyrosination by pharmacological treatment inhibited detrimental Ca^{2+} signals, protecting dystrophic hearts and skeletal muscles from damage [16].

As aforementioned, S-nitrosylation of RyR and its resulting calstabin2 depletion and Sarco-Endoplasmic Reticulum Calcium ATPase (SERCA-SR Ca^{2+}) leaks are related to skeletal and cardiac muscle weakness. Fauconnier et al. demonstrated that avoiding the depletion of calstabin2 and/or RyR S-nitrosylation by using RyR-specific "rycal" S107 and N-acetyl cysteine (NAC) respectively, prevented cardiac arrhythmias in DMD [30]. Those beneficial effects have been observed also in skeletal muscle, in which treatment with S107, enhancing the binding of calstabin-1 to nitrosylated RyR1, reduced SR Ca^{2+} leak and calpain activation, protecting skeletal muscle from the damage [32].

Yet another target for DMD treatment is nNOS and the resulting dysregulations in NO. NO deficiency leads to an increase of HDAC5 phosphorylation and an altered histone modification profile. Pharmacological intervention with NO donors regulates muscle differentiation and regeneration, triggering the nuclear localization of HDAC5 and changing histone modifications in DMD muscle [13, 78, 79].

NO alteration in dystrophic mice is also related to a reduction in HDAC2 S-nitrosylation, which represents one of the factors behind the deregulated gene expression in dystrophic mice [12].

All of these studies highlight the link between genetic defects in mdx mice and epigenetic modifications responsible for dystrophic pathogenesis. Therefore, different interventions are focusing on the regulation of epigenetic landscape in dystrophic mice. For example, HDAC inhibitors (HDACi) are emerging as a potent pharmacological treatment to counteract the progression of DMD [80–84]. These studies are contributing to better understand the varying mechanisms underlying the beneficial effects of pharmacological treatment with HDACi and pave the way for a recent, ongoing clinical trial with children with DMD [85–87].

5.6 Conclusions

Protein structures can have different properties due to post-translational modifications of their residues. Changes in protein folding capacity seem to exhibit deregulated pathways that are responsible for neurodegenerative pathogeneses. As highlighted in this chapter, PTM studies show great potential to better understand mechanisms behind neurodegenerative diseases. In particular, attention is focused on neuromuscular disorders with the intent of broadening knowledge on the crucial involvement of PTMs in these severe diseases and highlighting how these chemical modifications can be the basis of degenerative phenotypes.

In the last years, research studies have grown to investigate neurodegeneration using mass spectrometry (MS)-based proteomics. The combination of MS-based approaches with more diverse methods allows a study to go beyond simple PTM and protein identification, and is able to provide a profile of complex proteins to elucidate pathogenic mechanisms and discover effective biomarkers [88–90]. Importantly, the applications of proteomic technologies, which are able to identify the causative PTMs of many biological processes, are proving to be promising tools both to provide insight into the pathogenesis of diseases and to propose therapeutic strategies for PTM-associated diseases, moreover assisting to identify therapeutic targets for neurodegenerative diseases.

References

1. McDonald, C. M., et al. (1995). Profiles of neuromuscular diseases. Duchenne muscular dystrophy. *American Journal of Physical Medicine & Rehabilitation, 74*, S70–S92.
2. Emery, A. E. H. (2002). The muscular dystrophies. *Lancet (London, England), 359*, 687–695.
3. Himeda, C. L., & Jones, P. L. (2019). The genetics and epigenetics of Facioscapulohumeral muscular dystrophy. *Annual Review of Genomics and Human Genetics, 20*, 265–291.
4. Shieh, P. B. (2013). Muscular dystrophies and other genetic myopathies. *Neurologic Clinics, 31*, 1009–1029.

5. Carter, J. C., Sheehan, D. W., Prochoroff, A., & Birnkrant, D. J. (2018). Muscular dystrophies. *Clinics in Chest Medicine, 39*, 377–389.
6. Jin, Y., Diffee, G. M., Colman, R. J., Anderson, R. M., & Ge, Y. (2019). Top-down mass spectrometry of Sarcomeric protein post-translational modifications from non-human primate skeletal muscle. *Journal of the American Society for Mass Spectrometry, 30*, 2460–2469.
7. Uversky, V. N. (2013). Intrinsic disorder-based protein interactions and their modulators. *Current Pharmaceutical Design, 19*, 4191–4213.
8. Ervasti, J. M., Ohlendieck, K., Kahl, S. D., Gaver, M. G., & Campbell, K. P. (1990). Deficiency of a glycoprotein component of the dystrophin complex in dystrophic muscle. *Nature, 345*, 315–319.
9. Brenman, J. E., Chao, D. S., Xia, H., Aldape, K., & Bredt, D. S. (1995). Nitric oxide synthase complexed with dystrophin and absent from skeletal muscle sarcolemma in Duchenne muscular dystrophy. *Cell, 82*, 743–752.
10. Grozdanovic, Z., & Baumgarten, H. G. (1999). Nitric oxide synthase in skeletal muscle fibers: A signaling component of the dystrophin-glycoprotein complex. *Histology and Histopathology, 14*, 243–256.
11. Nott, A., Watson, P. M., Robinson, J. D., Crepaldi, L., & Riccio, A. (2008). S-nitrosylation of histone deacetylase 2 induces chromatin remodelling in neurons. *Nature, 455*, 411–415.
12. Colussi, C., et al. (2008). HDAC2 blockade by nitric oxide and histone deacetylase inhibitors reveals a common target in Duchenne muscular dystrophy treatment. *Proceedings of the National Academy of Sciences of the United States of America, 105*, 19183–19187.
13. Colussi, C., et al. (2009). Nitric oxide deficiency determines global chromatin changes in Duchenne muscular dystrophy. *The FASEB Journal, 23*, 2131–2141.
14. Garnham, C. P., et al. (2015). Multivalent microtubule recognition by tubulin tyrosine ligase-like family Glutamylases. *Cell, 161*, 1112–1123.
15. Argaraña, C. E., Barra, H. S., & Caputto, R. (1978). Release of [14C]tyrosine from tubulinyl-[14C]tyrosine by brain extract. Separation of a carboxypeptidase from tubulin-tyrosine ligase. *Molecular and Cellular Biochemistry, 19*.
16. Kerr, J. P., et al. (2015). Detyrosinated microtubules modulate mechanotransduction in heart and skeletal muscle. *Nature Communications, 6*, 8526.
17. Khairallah, R. J., et al. (2012). Microtubules underlie dysfunction in Duchenne muscular dystrophy. *Science Signaling, 5*, ra56–ra56.
18. Belanto, J. J., et al. (2016). Independent variability of microtubule perturbations associated with dystrophinopathy. *Human Molecular Genetics, ddw318*. https://doi.org/10.1093/hmg/ddw318
19. Prosser, B. L., Ward, C. W., & Lederer, W. J. (2011). X-ROS signaling: Rapid Mechano-chemo transduction in heart. *Science, 333*, 1440–1445.
20. Capitanio, D., et al. (2020). Comparative proteomic analyses of Duchenne muscular dystrophy and Becker muscular dystrophy muscles: Changes contributing to preserve muscle function in Becker muscular dystrophy patients. *Journal of Cachexia, Sarcopenia and Muscle, 11*, 547–563.
21. Gundersen, G. G., Kalnoski, M. H., & Bulinski, J. C. (1984). Distinct populations of microtubules: Tyrosinated and nontyrosinated alpha tubulin are distributed differently in vivo. *Cell, 38*, 779–789.
22. Kreitzer, G., Liao, G., & Gundersen, G. G. (1999). Detyrosination of tubulin regulates the interaction of intermediate filaments with microtubules in vivo via a kinesin-dependent mechanism. *Molecular Biology of the Cell, 10*, 1105–1118.
23. Ward, C. W., Prosser, B. L., & Lederer, W. J. (2014). Mechanical stretch-induced activation of ROS/RNS signaling in striated muscle. *Antioxidants & Redox Signaling, 20*, 929–936.
24. Prosser, B. L., Khairallah, R. J., Ziman, A. P., Ward, C. W., & Lederer, W. J. (2013). X-ROS signaling in the heart and skeletal muscle: Stretch-dependent local ROS regulates [Ca2+]i. *Journal of Molecular and Cellular Cardiology, 58*, 172–181.

25. Webster, D. R., & Patrick, D. L. (2000). Beating rate of isolated neonatal cardiomyocytes is regulated by the stable microtubule subset. *The American Journal of Physiology-Heart and Circulatory Physiology, 278*, H1653–H1661.
26. Jung, C., Martins, A. S., Niggli, E., & Shirokova, N. (2008). Dystrophic cardiomyopathy: Amplification of cellular damage by Ca2+ signalling and reactive oxygen species-generating pathways. *Cardiovascular Research, 77*, 766–773.
27. Deconinck, N., & Dan, B. (2007). Pathophysiology of Duchenne muscular dystrophy: Current hypotheses. *Pediatric Neurology, 36*, 1–7.
28. Erickson, J. R., & Anderson, M. E. (2008). CaMKII and its role in cardiac arrhythmia. *Journal of Cardiovascular Electrophysiology, 19*, 1332–1336.
29. Kyrychenko, S., et al. (2013). Hierarchical accumulation of RyR post-translational modifications drives disease progression in dystrophic cardiomyopathy. *Cardiovascular Research, 97*, 666–675.
30. Fauconnier, J., et al. (2010). Leaky RyR2 trigger ventricular arrhythmias in Duchenne muscular dystrophy. *Proceedings of the National Academy of Sciences, 107*, 1559–1564.
31. Adamo, C. M., et al. (2010). Sildenafil reverses cardiac dysfunction in the mdx mouse model of Duchenne muscular dystrophy. *Proceedings of the National Academy of Sciences, 107*, 19079–19083.
32. Bellinger, A. M., et al. (2009). Hypernitrosylated ryanodine receptor calcium release channels are leaky in dystrophic muscle. *Nature Medicine, 15*, 325–330.
33. Muntoni, F., Brockington, M., Blake, D. J., Torelli, S., & Brown, S. C. (2002). Defective glycosylation in muscular dystrophy. *The Lancet, 360*, 1419–1421.
34. Heydemann, A., & McNally, E. M. (2007). Consequences of disrupting the dystrophin-sarcoglycan complex in cardiac and skeletal myopathy. *Trends in Cardiovascular Medicine, 17*, 55–59.
35. Ehmsen, J., Poon, E., & Davies, K. (2002). The dystrophin-associated protein complex. *Journal of Cell Science, 115*, 2801–2803.
36. Sancar, F., et al. (2011). The dystrophin-associated protein complex maintains muscle excitability by regulating ca $^{2+}$ –dependent K $^{+}$ (BK) channel localization. *The Journal of Biological Chemistry, 286*, 33501–33510.
37. Kamdar, F., & Garry, D. J. (2016). Dystrophin-deficient cardiomyopathy. *Journal of the American College of Cardiology, 67*, 2533–2546.
38. Quattrocelli, M., Spencer, M. J., & McNally, E. M. (2017). Outside in: The matrix as a modifier of muscular dystrophy. *Biochimica et Biophysica Acta (BBA) - Molecular Cell Research, 1864*, 572–579.
39. Guiraud, S., et al. (2015). The pathogenesis and therapy of muscular dystrophies. *Annual Review of Genomics and Human Genetics, 16*, 281–308.
40. Ibraghimov-Beskrovnaya, O., et al. (1992). Primary structure of dystrophin-associated glycoproteins linking dystrophin to the extracellular matrix. *Nature, 355*, 696–702.
41. Moore, C. J., & Hewitt, J. E. (2009). Dystroglycan glycosylation and muscular dystrophy. *Glycoconjugate Journal, 26*, 349–357.
42. Cohn, R. D. (2005). Dystroglycan: Important player in skeletal muscle and beyond. *Neuromuscular Disorders, 15*, 207–217.
43. Winder, S. J. (2001). The complexities of dystroglycan. *Trends in Biochemical Sciences, 26*, 118–124.
44. Blake, D. J., Weir, A., Newey, S. E., & Davies, K. E. (2002). Function and genetics of dystrophin and dystrophin-related proteins in muscle. *Physiological Reviews, 82*, 291–329.
45. Gee, S. H., et al. (1993). Laminin-binding protein 120 from brain is closely related to the dystrophin-associated glycoprotein, dystroglycan, and binds with high affinity to the major heparin binding domain of laminin. *The Journal of Biological Chemistry, 268*, 14972–14980.
46. Ervasti, J. M., Burwell, A. L., & Geissler, A. L. (1997). Tissue-specific heterogeneity in α-Dystroglycan Sialoglycosylation: Skeletal muscle α-Dystroglycan is a latent receptor for

Vicia villosa agglutinin B $_4$ masked by sialic acid modification. *The Journal of Biological Chemistry, 272*, 22315–22321.
47. Brancaccio, A., Schulthess, T., Gesemann, M., & Engel, J. (1995). Electron microscopic evidence for a mucin-like region in chick muscle α-dystroglycan. *FEBS Letters, 368*, 139–142.
48. Brancaccio, A., Schulthess, T., Gesemann, M., & Engel, J. (1997). The N-terminal region of alpha-Dystroglycan is an autonomous globular domain. *European Journal of Biochemistry, 246*, 166–172.
49. Yoshida, A., et al. (2001). Muscular dystrophy and neuronal migration disorder caused by mutations in a glycosyltransferase, POMGnT1. *Developmental Cell, 1*, 717–724.
50. de Bernabé, D. B.-V., et al. (2009). Loss of alpha-dystroglycan laminin binding in epithelium-derived cancers is caused by silencing of LARGE. *The Journal of Biological Chemistry, 284*, 11279–11284.
51. Michele, D. E., et al. (2002). Post-translational disruption of dystroglycan–ligand interactions in congenital muscular dystrophies. *Nature, 418*, 417–421.
52. Willer, T., et al. (2012). ISPD loss-of-function mutations disrupt dystroglycan O-mannosylation and cause Walker-Warburg syndrome. *Nature Genetics, 44*, 575–580.
53. Sharpe, K. M., Premsukh, M. D., & Townsend, D. (2013). Alterations of dystrophin-associated glycoproteins in the heart lacking dystrophin or dystrophin and utrophin. *Journal of Muscle Research and Cell Motility, 34*, 395–405.
54. Chandrasekharan, K. & Martin, P. T. Genetic Defects in Muscular Dystrophy. in *Methods in Enzymology* vol. 479 291–322 (Elsevier, 2010).
55. Martin, P. T., et al. (2009). Overexpression of *Galgt2* in skeletal muscle prevents injury resulting from eccentric contractions in both mdx and wild-type mice. *American Journal of Physiology-Cell Physiology, 296*, C476–C488.
56. Minetti, C., Bado, M., Morreale, G., Pedemont, M., & Cordone, G. (1996). Disruption of muscle basal lamina in congenital muscular dystrophy with merosin deficiency. *Neurology, 46*, 1354–1354.
57. Beltrán-Valero de Bernabé, D., et al. (2002). Mutations in the O-mannosyltransferase gene POMT1 give rise to the severe neuronal migration disorder Walker-Warburg syndrome. *American Journal of Human Genetics, 71*, 1033–1043.
58. van Reeuwijk, J. (2005). POMT2 mutations cause -dystroglycan hypoglycosylation and Walker-Warburg syndrome. *Journal of Medical Genetics, 42*, 907–912.
59. Manya, H., et al. (2004). Demonstration of mammalian protein O-mannosyltransferase activity: Coexpression of POMT1 and POMT2 required for enzymatic activity. *Proceedings of the National Academy of Sciences, 101*, 500–505.
60. Sabatelli, P. et al. Extracellular matrix and nuclear abnormalities in skeletal muscle of a patient with Walker–Warburg syndrome caused by POMT1 mutation. Biochimica et Biophysica Acta - Molecular Basis of Disease 1638, 57–62 (2003).
61. Manya, H., et al. (2003). Loss-of-function of an N-acetylglucosaminyltransferase, POMGnT1, in muscle–eye–brain disease. *Biochemical and Biophysical Research Communications, 306*, 93–97.
62. Kano, H., et al. (2002). Deficiency of α-Dystroglycan in muscle–eye–brain disease. *Biochemical and Biophysical Research Communications, 291*, 1283–1286.
63. Kobayashi, K., et al. (1998). An ancient retrotransposal insertion causes Fukuyama-type congenital muscular dystrophy. *Nature, 394*, 388–392.
64. Aravind, L., & Koonin, E. V. (1999). The fukutin protein family – Predicted enzymes modifying cell-surface molecules. *Current Biology, 9*, R836–R837.
65. Brockington, M. (2001). Mutations in the fukutin-related protein gene (FKRP) identify limb girdle muscular dystrophy 2I as a milder allelic variant of congenital muscular dystrophy MDC1C. *Human Molecular Genetics, 10*, 2851–2859.
66. Brockington, M., et al. (2001). Mutations in the Fukutin-related protein gene (FKRP) cause a form of congenital muscular dystrophy with secondary Laminin α2 deficiency and abnormal glycosylation of α-Dystroglycan. *American Journal of Human Genetics, 69*, 1198–1209.

67. Inamori, K., et al. (2013). Xylosyl- and glucuronyltransferase functions of LARGE in α-dystroglycan modification are conserved in LARGE2. *Glycobiology, 23*, 295–302.
68. Longman, C. (2003). Mutations in the human LARGE gene cause MDC1D, a novel form of congenital muscular dystrophy with severe mental retardation and abnormal glycosylation of -dystroglycan. *Human Molecular Genetics, 12*, 2853–2861.
69. Grewal, P. K., Holzfeind, P. J., Bittner, R. E., & Hewitt, J. E. (2001). Mutant glycosyltransferase and altered glycosylation of α-dystroglycan in the myodystrophy mouse. *Nature Genetics, 28*, 151–154.
70. Reed, P. W., Mathews, K. D., Mills, K. A., & Bloch, R. J. (2004). The sarcolemma in the Large(myd) mouse. *Muscle & Nerve, 30*, 585–595.
71. Goddeeris, M. M., et al. (2013). LARGE glycans on dystroglycan function as a tunable matrix scaffold to prevent dystrophy. *Nature, 503*, 136–140.
72. Zhang, Z., Zhang, P., & Hu, H. (2011). LARGE expression augments the glycosylation of glycoproteins in addition to α-Dystroglycan conferring Laminin binding. *PLoS One, 6*, e19080.
73. Roscioli, T., et al. (2012). Mutations in ISPD cause Walker-Warburg syndrome and defective glycosylation of α-dystroglycan. *Nature Genetics, 44*, 581–585.
74. Cirak, S., et al. (2013). ISPD gene mutations are a common cause of congenital and limb-girdle muscular dystrophies. *Brain: A Journal of Neurology, 136*, 269–281.
75. Kanagawa, M., et al. (2016). Identification of a post-translational modification with Ribitol-phosphate and its defect in muscular dystrophy. *Cell Reports, 14*, 2209–2223.
76. Nguyen, H. H., Jayasinha, V., Xia, B., Hoyte, K., & Martin, P. T. (2002). Overexpression of the cytotoxic T cell GalNAc transferase in skeletal muscle inhibits muscular dystrophy in mdx mice. *Proceedings of the National Academy of Sciences, 99*, 5616–5621.
77. Xu, R., Chandrasekharan, K., Yoon, J. H., Camboni, M., & Martin, P. T. (2007). Overexpression of the cytotoxic T cell (CT) carbohydrate inhibits muscular dystrophy in the dyW mouse model of congenital muscular dystrophy 1A. *The American Journal of Pathology, 171*, 181–199.
78. Brunelli, S., et al. (2007). Nitric oxide release combined with nonsteroidal antiinflammatory activity prevents muscular dystrophy pathology and enhances stem cell therapy. *Proceedings of the National Academy of Sciences of the United States of America, 104*, 264–269.
79. Pisconti, A., et al. (2013). Follistatin induction by nitric oxide through cyclic GMP: A tightly regulated signaling pathway that controls myoblast fusion. *The Journal of Cell Biology, 200*, 359–359.
80. Minetti, G. C., et al. (2006). Functional and morphological recovery of dystrophic muscles in mice treated with deacetylase inhibitors. *Nature Medicine, 12*, 1147–1150.
81. Consalvi, S., et al. (2013). Preclinical studies in the mdx mouse model of duchenne muscular dystrophy with the histone deacetylase inhibitor givinostat. *Molecular Medicine (Cambridge, Mass.), 19*, 79–87.
82. Mozzetta, C., et al. (2013). Fibroadipogenic progenitors mediate the ability of HDAC inhibitors to promote regeneration in dystrophic muscles of young, but not old mdx mice. *EMBO Molecular Medicine, 5*, 626–639.
83. Consalvi, S., Saccone, V., & Mozzetta, C. (2014). Histone deacetylase inhibitors: A potential epigenetic treatment for Duchenne muscular dystrophy. *Epigenomics, 6*, 547–560.
84. Sandoná, M., Consalvi, S., Tucciarone, L., Puri, P. L., & Saccone, V. (2016). HDAC inhibitors for muscular dystrophies: Progress and prospects. *Expert Opinion on Orphan Drugs, 4*, 125–127.
85. Saccone, V., et al. (2014). HDAC-regulated myomiRs control BAF60 variant exchange and direct the functional phenotype of fibro-adipogenic progenitors in dystrophic muscles. *Genes & Development, 28*, 841–857.
86. Sandonà, M., et al. (2020). HDAC inhibitors tune miRNAs in extracellular vesicles of dystrophic muscle-resident mesenchymal cells. *EMBO Reports, 21*, e50863.
87. Bettica, P., et al. (2016). Histological effects of givinostat in boys with Duchenne muscular dystrophy. *Neuromuscular Disorders NMD, 26*, 643–649.

88. Zhang, J., Keene, C. D., Pan, C., Montine, K. S., & Montine, T. J. (2008). Proteomics of human neurodegenerative diseases. *Journal of Neuropathology and Experimental Neurology, 67*, 923–932.
89. Lovestone, S., et al. (2007). Proteomics of Alzheimer's disease: Understanding mechanisms and seeking biomarkers. *Expert Review of Proteomics, 4*, 227–238.
90. Schulenborg, T., et al. (2006). Proteomics in neurodegeneration – Disease driven approaches. *Journal of Neural Transmission, 113*, 1055–1073.

Chapter 6
Post-translational Modifications in Parkinson's Disease

André Saraiva Leão Marcelo Antunes

Abstract Parkinson's disease (PD) is a complex neurodegenerative disorder characterised by progressive degeneration of the dopaminergic neurons in the substantia nigra leading to severe motor complications. The etiology of the disease is unknown with its sporadic form accounting for 90% of cases. To date, over 20 genes have been identified as the cause of the inherited form of PD, many of them linked to the protein alpha-synuclein and mitochondrial function. Post-translational modifications of proteins allow cells to dynamically control signalling networks and diversify protein functions. This chapter will discuss briefly the main types of post-translational modifications, how to study them and how they affect proteins involved in PD.

Keywords Parkinson's disease · Post translation modifications · Mass spectrometry

6.1 Parkinson's Disease

In the beginning of the nineteenth century, James Parkinson described a "shaky palsy" that would later on be named Parkinson's Disease (PD) [1]. This progressive disorder is characterized by motor symptoms, which include akinesia (loss or impairment of voluntary movement), bradykinesia (slowness), cogwheel rigidity, resting tremor and postural instability. Furthermore, affected people may experience a broad range of non-motor symptoms such as hallucinations, depression, pain, apathy, sexual difficulties, visual impairment, bowel incontinence and insomnia, all of which still remain under-recognised and, consequently, under-treated [2]. PD is the second most common neurodegenerative disorder, affecting about 1–2% of people above the age of 55 with a higher incidence expected with the increasing life expectancy worldwide [3]. The etiology of PD remains unknown nearly two

A. S. L. M. Antunes (✉)
Brazilian Biosciences National Laboratory (LNBio), Brazilian Center for Research in Energy and Materials (CNPEM), São Paulo, Brazil

V. Corasolla Carregari (ed.), *Understanding PTMs in Neurodegenerative Diseases*, Advances in Experimental Medicine and Biology 1382,
https://doi.org/10.1007/978-3-031-05460-0_6

centuries after it was described. It is now clear that it is a complex multifactorial disorder with contributions from environmental and genetic factors. Genetic roots account for 10% of the cases, while the other 90% are idiopathic (also known as sporadic), with age being the major risk factor. The pathophysiology of PD is the progressive death of dopaminergic neurons from the nigrostiatal pathway that eventually leads to a dopamine deficiency that culminates in the manifestation of motor complications. Neuronal death in PD patients has been associated with the presence of intracellular inclusions of alpha-synuclein aggregates known as Lewy Bodies (LB) and Lewy Neurite (LN)s as well as abnormal mitochondrial function.

6.1.1 Alpha-Synuclein Aggregation and Parkinson's Disease

Alpha-synuclein, a cytoplasmic protein mainly present in presynaptic terminals, is a major component of LB and LN. The function of alpha-synuclein is not yet fully understood but it is thought to play a role in lipid binding and vesicle dynamics. It can form toxic insoluble cytoplasmic aggregates that spread from neuron to neuron in a prion like fashion, eventually reaching different interconnected areas in the nervous system. In the brain, abnormal alpha-synuclein deposition generally begins in the olfactory bulb and spreads to the medulla oblongata and the pontine tegumentum before reaching the amygdala and substantia nigra, where neuronal death leads to the appearance of classical motor symptoms of PD. From the midbrain, alpha-synuclein aggregation spreads to cortical areas, where it is associated with the occurrence of non-motor symptoms (Braak et al.). PD is considered to be synucleinopathy with distinct alpha-synuclein spreading patterns between genetic and sporadic forms, raising the question whether all forms of PD are the same [4]. Overall, there is strong evidence that suggests that the aberrant alpha-synuclein deposition is the driving force in PD pathogenesis [5]. To date, several alpha-synuclein post-translational modifications (PTMs) have been identified that seem to correlate with pathology formation.

6.1.2 Mitochondrial Dysfunction and Parkinson's Disease

Abnormal mitochondrial function and increased oxidative stress have also been implicated in the pathogenesis of PD. The first link appeared several decades ago when some heroin users developed PD-like symptoms after injecting an impure drug batch. It was established later that MPTP, a contaminant preferentially uptaken by dopaminergic neurons, was the underlying cause of the symptoms. MPTP acts by inhibiting the mitochondrial complex I, thus leading to oxidative stress followed by cell death [6]. Further evidence providing a link between abnormal mitochondrial function and PD pathology came from the identification of genes that are responsible for the less common genetic forms of PD. Mutations in the SNCA (alpha-synuclein) and LRRK2 (leucine-rich repeat kinase 2) genes are responsible

for the autosomal dominant forms of PD, whilst mutations in the genes encoding parkin, PINK1 (PTEN-induced putative kinase 1), HTRA2 (high temperature requirement protein A2) and DJ-1 mediate autosomal recessive forms of PD. These genes affect mitochondrial integrity in a direct or indirect manner.
Because sporadic and genetic forms of PD share important clinical, pathological and biochemical features, in particular the progressive loss of dopaminergic neurons in the substantia nigra, insight into the function of PD-associated gene products can help elucidate the primary mechanisms leading to neuronal death.

6.2 Post-translational Modifications

PTMs are chemical modifications to proteins that occur at any given time after they have been synthesized by ribosomes and they are to a great extent responsible for the complexity of the human proteome [7]. PTMs play a central role in cell division, signaling, protein localization, degradation, protein–protein interactions and enzymatic activity [8]. It is therefore not surprising that around 80% of mammalian proteins are subject to some kind of PTM [9]. Among the most common PTMs are phosphorylation, acetylation, glycosylation, and ubiquitination, which can occur indiviadually in a given protein or in conjunction with others [10]. PTMs play an elemental role in homeostasis and pathogenesis [11].

6.2.1 Phosphorylation

Protein phosphorylation is the most common form of PTM and it is responsible for translating virtually all kinds of extracellular signals into physiological effects by modulating the phosphorylation state of specific proteins inside the cells. Phosphorylation is a reversible modification orchestrated by a finely tuned balance between protein kinases and protein phosphatases. Protein kinases catalyse the transfer of a phosphate from ATP to the hydroxyl group of a target amino acid residue while protein phosphatases promote phosphate removal by means of hydrolysis [12]. Phosphate addition and removal occur mainly at serine (95%), and to a lesser extent on threonine (4%) and tyrosine (1%) residues [13]. Thus, an imbalance in this equilibrium leads to the hyper- or hypo-phosphorylation of target proteins.

6.2.2 Glycosylation

Glycosylation is one of the most complex PTMs and comprises the addition of oligosaccharides to proteins. It occurs in over 50% of human proteins lending them diverse biological functions [14]. Glycosylation occurs in the endoplasmic reticulum(ER)/Golgi compartment and can be divided into three main types:

N-linked and O-linked glycosylation, and glypiation [15]. Like protein phosphorylation, glycosylation is essential for cellular homeostasis and has been shown to play a role in pathology, including PD.

6.2.3 *Acetylation*

Acetylation consists in the reversible transfer of an acetyl group from the acetyl-coenzyme A to the ε-amino group of lysine residues side chain. Reactions are catalyzed by histone acetyl transferases (HATs) and histone deacetylases (HDACs). Despite owing their name to their best known substrates, HATs and HDACs have been shown to acetylate a number of proteins; their effects extend beyond transcriptional regulation and affect cellular metabolism and cell signaling [16]. Like phosphorylation, the balance between acetylation and deacetylation is crucial for cellular homeostasis and a dysregulated balance of these processes is linked to neurodegeneration [17].

6.2.4 *Ubiquitination*

Ubiquitination is a PTM that affects several cellular processes including protein intracellular localization, activation/inactivation of proteins, protein-protein interactions and protein degradation [18–20]. The Ubiquitin-Proteasome System (UPS) is the major cellular pathway for degradation of misfolded proteins and disorders in the UPS are related to intracellular protein aggregation, cytotoxicity and cell death [21]. As a matter of fact, neurodegenarative diseases, including PD, are characterized by the accumulation of aggregated misfolded proteins and inclusion bodies.
Ubiquitin is a 76 amino acid expressed in virtually all eukaryotic cells and ubiquitination is catalyzed by successive reactions that involve an ubiquitin-activating enzyme (E1), an ubiquitin-conjugating enzyme (E2) and an ubiquitin-protein ligase (E3) [22]. Together, these enzymes are responsible for the addition of one or several ubiquitin moieties to a lysine residue in the target protein. The PTM of a protein by means of mono-ubiquitination is responsible for the modulation of cellular events, whereas poly-ubiquitination, i.e. the addition of several ubiquitin moieties to previously conjugated ubiquitin molecules, leads to protein degradation by the UPS [23].

6.3 Mass Spectrometry in the Investigation of PTMs

Mass spectrometry is a powerful tool that allows scientists to gain detailed insight into disease-associated gene products and over the past two decades it has become increasingly relevant in studies of neurodegenerative diseases [24]. Mass spectrometry is a potent tool in the identification and characterization of PTMs [25] and improvements in its technical and analytical capabilities have contributed greatly to the identification of low stoichiometry PTMs in complex samples. There are four main mass spectrometry approaches for proteomic studies: bottom-up or shotgun, middle-down, top-down, and targeted proteomics. Shotgun or bottom-up mass spectrometry is the most commonly used approach in the investigation of complex physiopathological mechanisms. In this approach, proteins are extracted from a biological sample and cleaved at specific amino acid residues, thereby generating large numbers of peptides. These peptides undergo separation by liquid chromatography and are subsequently ionized and injected into a mass spectrometer, which generates spectra that allow for the quantitation and identification of the parent proteins. Middle-down, which generates larger peptides after cleavage, and top-down proteomics, which bypasses the protein cleavage step prior to mass spectrometry analysis, allow for increased coverage and better identification of PTMs [26, 27], however there are still some hurdles in the data analysis step of these approaches that make bottom-up proteomics the most used approach for PTM investigation. Despite being the method of choice, the bottom-up approach also faces some hurdles. PTMs such as phosphorylation or glycosylation lead to less efficient peptide ionization relative to their non-modified counterparts due to increased negative charge and acidity, thus, becoming more difficult to identify them due to ionic suppression.

To overcome this hurdle, methods have been developed that allow the phospho- and glycoproteome analysis by mass spectrometry. For phosphoproteome analysis, phosphopeptides are usually enriched either at the phosphoprotein or at the phosphopeptide level. Enrichment strategies are based on affinity chromatography that captures the negatively charged phosphate groups or by immunoprecipitation [28]. For glycoproteomic analysis, glycans are cleaved from glycoproteins, followed by purification steps prior to mass spectrometry analysis. This method, however, is suitable for glycoproteins containing only one glycosylation site, because it does not allow correlating glycan composition with different glycosylation sites on the peptide. In order to do such correlation, a glycosylation site-specific analysis must be performed, which is based on intact glycopeptides released by glycoprotein proteolysis [28, 29].

6.4 PTMs and PD

6.4.1 Phosphorylation

Alpha-synuclein is highly enriched in the presynaptic terminals of neurons and is natively unfolded. Abnormal aggregation and accumulation of alpha-synuclein is the histological hallmark of PD. To date, several PTMs have been identified as essential for alpha-synuclein adequate function clearance, including phosphorylation, ubiquitination and glycosylation [30, 31]. In LB, alpha-synuclein fibrils are commonly phosphorylated at S87, S129, Y39, Y125, Y133 and Y135 as well as highly ubiquitinated in its N terminus [32–35]. Some of these phosphorylations can accelerate alpha-synuclein aggregation and neuropathology in a transgenic mouse model of PD [36]. Under physiological conditions, around 4% of alpha-synuclein is phosphorylated, while it reaches up to 90% phosphorylation in LB, suggesting a role for alpha-synuclein PTM in the etiology of PD [37].

Mutations in the LRRK2 gene are the major cause of the autosomal-dominant form of familial PD. Overall, these mutations increase LRRK2 phosphorylation rate which in turn lead to the development of PD [38]. Using an unbiased phosphoproteomics approach in combination with genetically engineered mouse cells and a mixture of selective LRRK2 inhibitors, researchers have identified physiological substrates of LRRK2 and shed light into their contribution to the development of PD [39]. By using high stringency criteria, they identified the T73 phosphorylation site in the Rab10 GTPase. The Thr73 phosphorylation site was further validated in different Rabs containing this conserved threonine residue.

Mutations in *PARK6*, the coding gene of PINK1 are implicated with the autosomal recessive form of familial PD [40]. Using a phosphoproteomic approach, another study set out to identify PINK1-dependent phosphorylation targets in HEK293T cells [41] found Rab GTPases Rab8A, 8B and 13 to be phosphorylated at S111 in response to PINK1 activation. Further, Rab8A phosphorylation at S111 impairs its activation by its activator Rabin8, thus disrupting the Rab GTPase-mediated signalling. Rab GTPases play a central role in intracellular cargo transport and disruption of this signaling pathway may represent a pathophysiological mechanism of PD. Interestingly, unlike its well known targets ubiquitin and Parkin, PINK1 failed to phosphorylate the identified Rabs in vitro, suggesting the participation of a further kinase as an intermediate player.

Aberrant kinase function lead to an imbalanced phosphorylation of their substrates and LRRK2 and PINK1 are examples of how PTMs impact health and disease.

6.4.2 Ubiquitination

In addition to phosphorylation, ubiquitination has been shown to play a role in the development of PD. Both PTMs often occur together in signaling pathways. Phosphorylation of particular residues can regulate the addition of ubiquitin moieties by E3-ubiquitin ligases onto lysine residues of a substrate. In fact, PARKIN, an E3-Ubiqiutin ligase, acts together with PINK1 in an complex interplay between ubiquitination and phosphorylation that regulates mitochondrial quality control. In defective mitochondria, PINK1 accumulates at the outer mitochondria surface and leads to PARKIN translocation from the cytosol to mediate mitophagy [42, 43]. PARKIN translocation into the damaged mitochondria occurs upon its phosphorylation by PINK1, which also phosphorylates Ub conjugates, thereby enhancing PARKIN translocation [44]. Once in the mitochondria, PARKIN ubiquitinates and mediates the clearance of several mitochondrial proteins, including itself, thus leading to mitophagy [45, 46]. In summary, mitochondrial quality control by PINK1- and PARKIN demonstrates the intricate role of PTM in mitochondrial function, and consequently for the development of PD.

Further evidence of the importance of PTMs, more specifically of the ubiquitination/deubiquitination machinery in the etiology of PD is the fact that ubiquitin carboxyl-terminal hydrolase L1 (UCH-L1), a neuronal de-ubiquitinating enzyme is downregulated in early-onset PD patients and has been linked with the manifestation of idiopathic PD [47].

6.4.3 Glycosylation

Structural elucidation of glycoproteins still poses a technical challenge due to the inherent complexity and heterogeneity of the glycans attached to proteins. However, advances have allowed the identification of proteins in which glycosylation appears to play a major role in the development of PD [48]. For instance, one prospective study managed to characterise overlapping glycoproteins in the cerebrospinal fluid and postmortem brain tissue of PD patients using MALDI-TOF-TOF and identified a total of 283 and 394 non-redundant glycoproteins, respectively [49]. In addition, the glycoproteins neuronal pentraxin II and neuroserpin have been shown to be differentially regulated in PD [50, 51]. It is known that N-linked glycosylation is particularly prevalent in proteins destined for extracellular environments [52]. The impact of glycosylation of β4, a sodium channel that contains four potential N-linked glycosylation sites and whose expression has been correlated with PD pathogenesis, has been demonstrated in an immortalized neuronal cell line [53]. Furthermore, it has been shown that O-GlcNAcylation of alpha-synuclein can hinder its fibrillation [31].

6.5 Conclusion

Understanding of PTM diversity and its role in tissue homeostasis and disease has advanced immensely in the last decades. This has been partly driven by the development and refinement of tools such as mass spectrometry and biotechnology that allow us to better characterise macromolecules. Nonetheless, we are only beginning to understand the global complexity and regulation of PTMs. Challenges remain in the elucidation of the interplay between different PTMs and their combinatorial effects in health and disease states. In addition, the reversible nature of most PTMs and the relative low abundance of some of them creates challenges for researchers that attempt to capture the precise state of proteins in physiological and pathological conditions. Such challenges must be taken into account when choosing a method of study and development of methods that allow for in situ evaluation of PTMs remains a much-anticipated technological advancement.

References

1. Jost, W. H., & Reichmann, H. (2017). 'An essay on the shaking palsy' 200 years old. *Journal of Neural Transmission, 124*, 899–900.
2. Chaudhuri, K. R., Ray Chaudhuri, K., & Schapira, A. H. V. (2009). Non-motor symptoms of Parkinson's disease: dopaminergic pathophysiology and treatment. *The Lancet Neurology, 8*, 464–474.
3. de Lau, L. M. L., & Breteler, M. M. B. (2006). Epidemiology of Parkinson's disease. *Lancet Neurology, 5*, 525–535.
4. Doherty, K. M., et al. (2013). Parkin disease: a clinicopathologic entity? *JAMA Neurology, 70*, 571–579.
5. Stefanis, L. (2012). α-Synuclein in Parkinson's disease. *Cold Spring Harbor Perspectives in Medicine, 2*, a009399.
6. Subramaniam, S. R., & Chesselet, M.-F. (2013). Mitochondrial dysfunction and oxidative stress in Parkinson's disease. *Progress in Neurobiology, 106-107*, 17–32.
7. Ren, R.-J., Dammer, E. B., Wang, G., Seyfried, N. T., & Levey, A. I. (2014). Proteomics of protein post-translational modifications implicated in neurodegeneration. *Translational Neurodegeneration, 3*, 23.
8. Dewhurst, H. M., Choudhury, S., & Torres, M. P. (2015). Structural analysis of PTM hotspots (SAPH-ire)--A quantitative informatics method enabling the discovery of novel regulatory elements in protein families. *Molecular & Cellular Proteomics, 14*, 2285–2297.
9. Abou-Abbass, H., et al. (2016). Glycosylation and other PTMs alterations in neurodegenerative diseases: Current status and future role in neurotrauma. *Electrophoresis, 37*, 1549–1561.
10. Mann, M., & Jensen, O. N. (2003). Proteomic analysis of post-translational modifications. *Nature Biotechnology, 21*, 255–261.
11. Khoury, G. A., Baliban, R. C., & Floudas, C. A. (2011). Proteome-wide post-translational modification statistics: Frequency analysis and curation of the swiss-prot database. *Scientific Reports, 1*.
12. Hunter, T. (1995). Protein kinases and phosphatases: The yin and yang of protein phosphorylation and signaling. *Cell, 80*, 225–236.
13. Pawson, T., & Scott, J. D. (2005). Protein phosphorylation in signaling--50 years and counting. *Trends in Biochemical Sciences, 30*, 286–290.
14. Christiansen, M. N., et al. (2014). Cell surface protein glycosylation in cancer. *Proteomics, 14*, 525–546.
15. Spiro, R. G. (2002). Protein glycosylation: Nature, distribution, enzymatic formation, and disease implications of glycopeptide bonds. *Glycobiology, 12*, 43R–56R.
16. Choudhary, C., et al. (2009). Lysine acetylation targets protein complexes and co-regulates major cellular functions. *Science, 325*, 834–840.
17. Saha, R. N., & Pahan, K. (2006). HATs and HDACs in neurodegeneration: A tale of disconcerted acetylation homeostasis. *Cell Death and Differentiation, 13*, 539–550.
18. Schnell, J. D., & Hicke, L. (2003). Non-traditional functions of ubiquitin and ubiquitin-binding proteins. *The Journal of Biological Chemistry, 278*, 35857–35860.
19. Mukhopadhyay, D., & Riezman, H. (2007). Proteasome-independent functions of ubiquitin in endocytosis and signaling. *Science, 315*, 201–205.
20. Ciechanover, A. (2005). Intracellular protein degradation: From a vague idea thru the lysosome and the ubiquitin-proteasome system and onto human diseases and drug targeting. *Cell Death and Differentiation, 12*, 1178–1190.
21. Schwartz, A. L., & Ciechanover, A. (2009). Targeting proteins for destruction by the ubiquitin system: Implications for human pathobiology. *Annual Review of Pharmacology and Toxicology, 49*, 73–96.
22. Ciechanover, A., & Schwartz, A. L. (1998). The ubiquitin-proteasome pathway: The complexity and myriad functions of proteins death. *Proceedings of the National Academy of Sciences of the United States of America, 95*, 2727–2730.

23. Marx, J. (2002). Cell biology. Ubiquitin lives up to its name. *Science, 297*, 1792–1794.
24. Antunes, A. S. L. M., de Almeida, V., Crunfli, F., Carregari, V. C., & Martins-de-Souza, D. (2021). Proteomics for target identification in psychiatric and neurodegenerative disorders. *Advances in Experimental Medicine and Biology, 1286*, 251–264.
25. Jensen, O. N. (2004). Modification-specific proteomics: Characterization of post-translational modifications by mass spectrometry. *Current Opinion in Chemical Biology, 8*, 33–41.
26. Yu, D., Wang, Z., Cupp-Sutton, K. A., Liu, X., & Wu, S. (2019). Deep intact Proteoform characterization in human cell lysate using high-pH and low-pH reversed-phase liquid chromatography. *Journal of the American Society for Mass Spectrometry, 30*, 2502–2513.
27. Cupp-Sutton, K. A., & Wu, S. (2020). High-throughput quantitative top-down proteomics. *Molecular Omics, 16*, 91–99.
28. Fíla, J., & Honys, D. (2012). Enrichment techniques employed in phosphoproteomics. *Amino Acids, 43*, 1025–1047.
29. Zhu, R., Zacharias, L., Wooding, K. M., Peng, W., & Mechref, Y. (2017). Glycoprotein enrichment analytical techniques: Advantages and disadvantages. *Methods in Enzymology, 585*, 397–429.
30. Theillet, F.-X., et al. (2016). Structural disorder of monomeric α-synuclein persists in mammalian cells. *Nature, 530*, 45–50.
31. Levine, P. M., et al. (2019). α-Synuclein O-GlcNAcylation alters aggregation and toxicity, revealing certain residues as potential inhibitors of Parkinson's disease. *Proceedings of the National Academy of Sciences of the United States of America, 116*, 1511–1519.
32. Fujiwara, H., et al. (2002). alpha-Synuclein is phosphorylated in synucleinopathy lesions. *Nature Cell Biology, 4*, 160–164.
33. Paleologou, K. E., et al. (2010). Phosphorylation at S87 is enhanced in synucleinopathies, inhibits alpha-synuclein oligomerization, and influences synuclein-membrane interactions. *The Journal of Neuroscience, 30*, 3184–3198.
34. Negro, A., Brunati, A. M., Donella-Deana, A., Massimino, M. L., & Pinna, L. A. (2002). Multiple phosphorylation of alpha-synuclein by protein tyrosine kinase Syk prevents eosin-induced aggregation. *The FASEB Journal, 16*, 210–212.
35. Zhao, K., et al. (2020). Parkinson's disease-related phosphorylation at Tyr39 rearranges α-synuclein amyloid fibril structure revealed by cryo-EM. *Proceedings of the National Academy of Sciences of the United States of America, 117*, 20305–20315.
36. Dikiy, I., et al. (2016). Semisynthetic and in Vitro Phosphorylation of Alpha-Synuclein at Y39 Promotes Functional Partly Helical Membrane-Bound States Resembling Those Induced by PD Mutations. *ACS Chemical Biology, 11*, 2428–2437.
37. Anderson, J. P., et al. (2006). Phosphorylation of Ser-129 is the dominant pathological modification of alpha-synuclein in familial and sporadic Lewy body disease. *The Journal of Biological Chemistry, 281*, 29739–29752.
38. Zimprich, A., et al. (2004). Mutations in LRRK2 cause autosomal-dominant parkinsonism with pleomorphic pathology. *Neuron, 44*, 601–607.
39. Steger, M., et al. (2016). Phosphoproteomics reveals that Parkinson's disease kinase LRRK2 regulates a subset of Rab GTPases. *eLife, 5*.
40. Geisler, S., et al. (2010). The PINK1/Parkin-mediated mitophagy is compromised by PD-associated mutations. *Autophagy, 6*, 871–878.
41. Lai, Y.-C., et al. (2015). Phosphoproteomic screening identifies Rab GTPases as novel downstream targets of PINK1. *The EMBO Journal, 34*, 2840–2861.
42. Narendra, D. P., et al. (2010). PINK1 is selectively stabilized on impaired mitochondria to activate Parkin. *PLoS Biology, 8*, e1000298.
43. Greene, A. W., et al. (2012). Mitochondrial processing peptidase regulates PINK1 processing, import and Parkin recruitment. *EMBO Reports, 13*, 378–385.
44. Durcan, T. M., & Fon, E. A. (2015). The three 'P's of mitophagy: PARKIN, PINK1, and post-translational modifications. *Genes & Development, 29*, 989–999.

45. Sarraf, S. A., et al. (2013). Landscape of the PARKIN-dependent ubiquitylome in response to mitochondrial depolarization. *Nature, 496*, 372–376.
46. Chaugule, V. K., et al. (2011). Autoregulation of Parkin activity through its ubiquitin-like domain. *The EMBO Journal, 30*, 2853–2867.
47. Choi, J., et al. (2004). Oxidative modifications and down-regulation of ubiquitin carboxyl-terminal hydrolase L1 associated with idiopathic Parkinson's and Alzheimer's diseases. *The Journal of Biological Chemistry, 279*, 13256–13264.
48. Sihlbom, C., Davidsson, P., Sjögren, M., Wahlund, L.-O., & Nilsson, C. L. (2008). Structural and quantitative comparison of cerebrospinal fluid glycoproteins in Alzheimer's disease patients and healthy individuals. *Neurochemical Research, 33*, 1332–1340.
49. Hwang, H., et al. (2010). Glycoproteomics in neurodegenerative diseases. *Mass Spectrometry Reviews, 29*, 79–125.
50. Moran, L. B., et al. (2008). Neuronal pentraxin II is highly upregulated in Parkinson's disease and a novel component of Lewy bodies. *Acta Neuropathologica, 115*, 471–478.
51. Pisani, A., Centonze, D., Bernardi, G., & Calabresi, P. (2005). Striatal synaptic plasticity: Implications for motor learning and Parkinson's disease. *Movement Disorders, 20*, 395–402.
52. Roth, J. (2002). Protein N-glycosylation along the secretory pathway: Relationship to organelle topography and function, protein quality control, and cell interactions. *Chemical Reviews, 102*, 285–303.
53. Zhou, T.-T., Zhang, Z.-W., Liu, J., Zhang, J.-P., & Jiao, B.-H. (2012). Glycosylation of the sodium channel β4 subunit is developmentally regulated and involves in neuritic degeneration. *International Journal of Biological Sciences, 8*, 630–639.

Chapter 7
Histone Modifications in Neurological Disorders

Bradley J. Smith and Victor Corasolla Carregari

Abstract Post-translational modifications (PTMs) have a strong impact on many proteins across all kingdoms of life, affecting multiple functional and chemical properties of their protein recipients. With increasing knowledge about their functions, targets, and biological effects, dysregulations in PTMs have been implicated in various dysfunctions and diseases. One such target are histones, which compose the majority of the protein component of chromatin and the modulation of the 30+ PTMs that are known to affect them can have profound effects on chromatin state, gene expression, and DNA repair. In this review, the histone targets of PTMs are compiled in the context of neurological disorders, highlighting their specific biological roles and any previously implicated dysregulations in several classes of brain disease. Better understanding the pathogenic dysregulations of PTMs in such disorders can help to better understand their causes, as well as open doors to new possibilities for biomarkers and therapeutic targets.

Keywords Histones · Post-translational modifications · Psychiatric disorders · Neurodevelopmental disorders

7.1 Histones and Their Modifications

Within the nucleus of each human cell, approximately three billion base pairs are carefully compacted and twisted, tucking away some regions of the genome to the edges of the nucleus, while others are given center stage. The accessibility of these base pairs to other proteins is a vital part of transcriptional regulation, thereby making the regulation of DNA storage crucial to the proper functioning of cells. To compact the genetic material more efficiently, as well as provide a regulatory

---

B. J. Smith (✉) · V. C. Carregari
Laboratory of Neuroproteomics, Department of Biochemistry and Tissue Biology, Institute of Biology, University of Campinas (UNICAMP), Campinas, Brazil

V. Corasolla Carregari (ed.), *Understanding PTMs in Neurodegenerative Diseases*, Advances in Experimental Medicine and Biology 1382,
https://doi.org/10.1007/978-3-031-05460-0_7

scaffold, 147 base pairs of DNA are wrapped twice around cylindrical octameric protein complexes called histones. DNA can then be more tightly compacted or more loosely compacted as a way to encourage or discourage transcription, as well as to affect other processes like apoptosis, DNA repair, and DNA replication and mitosis.

This regulation is performed largely by post-translational modifications of histone subunits, making them a focus of study for decades. Lysine, arginine, and serine residues are some of the earliest known targets for modification by acetylation, methylation, phosphorylation, and ubiquitylation. Each modification can have multiple regulatory enzymes and can also function as an anchor for protein complexes [1], further increasing regulatory complexity. Over the past 20 years, several other modifications have been discovered with varying degrees of knowledge about their function and mechanisms of addition and removal (see Table 7.1). Most histone modifications often either directly modulate gene transcription through interactions with other protein partners or indirectly affect gene expression by affecting chromatin density, and some do both. Many modifications are also site-specific with regulatory enzymes performing PTM addition and/or removal, sometimes in response to specific cellular conditions. Due to their unique role in gene expression, histones are an important target when studying multifactorial psychiatric disorders and how to understand, prevent, and treat them.

7.2 Investigating Histone PTMs

When studying the levels of PTMs on histones, various biological questions can be answered by modifying the technique used. In rare modifications or in PTMs that respond directly to cellular conditions (e.g. non-enzymatically regulated), overall levels of modification may offer interesting results. This type of investigation has the simplest sample preparation, as it doesn't differ from most methods of sample preparation for shotgun proteomics. With newer instruments and their high resolution and sensitivity, it is possible to measure histone peptides and their modifications. Less sensitive instruments, however, may benefit from sample pre-processing such as abundant protein depletion or immuno-based pulldown of histones, which also enables studying histone interactomes. Crosslinking reagents can help to study the proteins and protein clusters that interact with modified histones.

Due to the large number of histone modifications and their sites, some experiments may benefit from an investigation into the co-expression of PTMs. In this case top-down proteomics must be employed, in which proteins are first purified by chromatographic separation before being directly ionized and submitted to MS/MS analysis. This method differs from bottom-up proteomics, in which a mixed protein extract is digested before ionization and analysis, resulting in a higher number of identified proteins at the cost of losing all co-expression and degradation data. Lastly, it is also possible to investigate the protein core as well as the DNA sequences wrapped around them. Using antibodies against modifications independently of

Table 7.1 A compilation of known histone modifications, their target amino acid(s), and any known effects of these modifications on chromatin state, gene expression, or biological function

Modification	Target residues	Biological effects
Acetylation	K, T	Transcriptional activation, histone deposition, DNA repair, euchromatin promotion
Benzoylation	K	Transcriptional activation*
Biotinylation	K	Cell proliferation, gene silencing, DNA repair
Butyrylation	K	Transcriptional activation*
Carbonylation	K	Cell proliferation, stress response
Crotonylation	K	Localization to promoters and enhancers
Deimination	R (to citrulline)	Transcription regulation
(N-)Formylation	K	Potentially pathogenic
Glutarylation	K	Chromatin structure, DNA repair
Glutathionylation	C	Opens chromatin structure
Hydroxylation	Y	Transcription regulation*, chromosomal rearrangement*
Isobutyrylation	K	Transcriptional activation
Isomerization	P (cis to trans)	Transcription regulation
Lactylation	K	Transcription regulation
Long fatty-chain acylation	K, S	Nuclear organization*, chromatin state*
Malonylation	K	Chromatin state*
Methylation	K (me1, me2, me3) R (me1, me2a, me2s)	Transcriptional silencing, repression, and elongation, (less commonly, activation); euchromatin and heterochromatin formation; DNA methylation
O-glcNAcylation	S, T	Cell cycle and heat shock, modification crosstalk
Phosphorylation	S, T	Transcription regulation, DNA repair, chromatin condensation
polyADP-ribosylation	E	Transcription regulation
Proprionylation	K	Transcriptional activation
Serotonylation	Q	Euchromatin formation*
Succinylation	K	Chromatin state*
SUMOylation	K	Transcription regulation
Ubiquitylation (UFMylation, ISG15ylation, NEDDylation)	K	Transcription regulation, DNA repair
β-hydroxybutyrylation	K	Transcriptional activation at promoters
2-hydroxyisobutyrylation	K	Transcriptional activation

Functions marked with an * are only suggested and not yet definitively proven 65–90

their flanking sequences, called pan-antibodies, it is possible to perform an immunoaffinity pulldown for modified histones, a modification of a technique called chromatin immunoprecipitation (ChIP). This step is performed after using a DNase to cleave all exposed DNA regions; and the DNA regions protected by histones can then be later sequenced, in a modified ChIP-seq protocol. In this way, the segments of the genome that interact with modified and unmodified histones can be identified.

7.3 Specific Roles in Neurological Disorders

Due to the important regulatory role that histone modifications can have on transcription and overall gene expression, it is no surprise that histone PTMs have been studied and implicated in the pathogenesis of various diseases, including neurological disorders. Understanding the differences in histone modifications for a given disease provides insight not only into the etiology of the disorder, but also highlights new lines of study to better identify, treat, and even prevent it altogether.

7.3.1 Alzheimer's, Parkinson's, and Huntington's Diseases

The biological characteristics of Alzheimer's disease (ALZ) are well known, and many advances have been made in determining proteins that lead to its pathology. Nonetheless, a panoramic view of the causes of these dysregulations is still unknown. Multiple factors contribute to the development of this disease, though the underlying connections between them remain hidden. In one study, histone acetylation levels, as well as histone levels overall, were found to be increased in ALZ [2]. These increases were found in areas that are associated with both the pathology of the disease (amyloid and tau) and ubiquitination pathways, leading to the hypothesis that protein degradation and turnover may be reduced in ALZ, leading to cell death. In another study, overall histone H3 phosphorylation was also increased [3], further supporting this hypothesis. Another particular gene of interest in ALZ, ankyrin 1 (ANK1), was targeted in a study using ChIP (chromatin immunoprecipitation) paired with qPCR, and the authors found that DNA methylation of that gene was associated with trimethylation of nearby histones [4]. Both factors lead to a reduced expression of their target gene, suggesting a potential therapeutic target for future studies.

Some correlations have been made between Parkinson's Disease (PD) and histone modifications. For example, alpha-synuclein, a protein found to have increased expression in PD, is known to inhibit histone acetylation [5] and some cases of PD have decreased histone acetylation in correlation with higher alpha-synuclein levels [6]. Dopamine depletion seen in some rodent and primate PD models has been associated with histone methylation, and the administration of a dopamine precursor partially improves symptoms and reduces some histone acetylation sites [7]. Both histone deacetylase inhibition and demethylase inhibition have been studied as treatment options for PD for neuroprotection [8, 9].

As reviewed by Gapp et al., histone acetylation, methylation, and ubiquitination have all been implicated in Huntington's Disease (HD) [10]. Later, Bassi et al. reviewed specific mechanisms of histone modification homeostasis in HD [11], also including other epigenetic changes and post-translational modifications of proteins. Bringing these findings slightly closer to treatment options, inhibiting histone deacetylases in mice has been shown to protect against metabolic dysfunctions in HD [12] and a histone deacetylase 3 inhibitor was able to prevent deficits in cognition and reduce CAG repeats in a mouse model [13]. The inhibition of HDACs 1 and 3 has also been shown to improve motor deficits in a mouse model of HD, offering yet another possible treatment option for future study [14].

7.4 Aging and Cognitive Decline

Age has also been documented to affect histone modifications, which has led to hypotheses about how these age-related changes may contribute to a decline in brain health, and subsequently, mental health [15]. This is seen, for example, in a loss of histone acetylation at the promoter regions of important genes for neurological functioning, including neurotransmission pathways and mitochondrial function [16]. In mice, epigenetic signatures associated with memory capacity and cognitive function were absent in aged mice, [17] and restoring decreased acetylation pathways in aged mice has been shown to reverse deficits in memory formation [18]. Modulating histone modifications has also been studied to treat age-associated cognitive decline, dementia, and late-onset Alzheimer's disease [15, 19, 20], though only small steps have been made towards understanding how to precisely and carefully make such modulations [21–23].

7.5 Cognition, Memory, and Neurodevelopment

Histone acetylation has long been associated with memory formation, particularly long-term memory [24]. Since this discovery, multiple sites of histone acetylation have been associated to varying degrees with memory function and cognition [25–27]. As reviewed by Fischer, conditions with cognitive symptoms or origins have also been implicated, such as Rubenstein-Taybi syndrome, post-traumatic stress disorder, schizophrenia, drug abuse, Alzheimer's disease, and Fragile X syndrome [28]. Various classes of histone deacetylases (HDACs) have also been extensively studied in relation to cognition, learning, memory, and neurodevelopment [29], and in one comprehensive review, a list of HDACs was compiled in relation to beneficial and harmful effects in Alzheimer's disease, depression, stress, schizophrenia, addiction, and autism spectrum disorders [30]. The biological effects of HDACs as well as histone acetyltransferases (HATs) were also compiled in relation to brain development and neurodevelopmental disorders [31].

Mutations in genes that code for regulatory enzymes of histone lysine methylation have been implicated in autism spectrum disorder (ASD) [32], though data have been contradictory [33], suggesting that there is yet much to learn about the correlation between epigenetics and a predisposition to ASD [34]. Neural tube defects occurring during neurodevelopment have been associated with a folate deficiency-induced lack of histone dimethylation, preventing proper protein interaction and subsequently negatively affecting gene expression [35]. X-linked mental retardation has also seen a correlation with aberrant histone methylation at important gene loci [36].

Development and specifically neurodevelopment also heavily rely on histone modifications to control gene expression [31, 37, 38]. With carefully modulated gene expression patterns that occur throughout the body sourced by sex chromosomes, a person's sex during development has been shown to have an interplay with histone modifications early during neurodevelopment [39–42] as well as later, during puberty [43]. This has been hypothesized to lead to some of possible sex-based differences in brain function [44] and therefore neurodevelopmental disorders. In fact, the proteome of schizophrenia in particular has been shown to differ between male and female patients [45]. Singh et al. have compiled the mechanisms and resulting biological changes of sex-related epigenetics in neuropsychiatric disorders in a comprehensive review [46].

7.6 Schizophrenia and Bipolar Disorder

The overall landscape of histone acetylation and trimethylation was mapped in the human frontal lobe, which showed that risk variants for schizophrenia, depression, and neuroticism were highly represented in neurons [47]. Obtaining histone modification maps for different cell types and regions in this way holds great potential to link genetic risk variants with the cells and tissues that are most affected by a disorder.

The etiology of schizophrenia (SCZ) has been linked with histone modification dysregulation at a genetic level using GWAS data, though no particular modification or site was determined in that report [48]. That being said, patients with SCZ and bipolar disorder (BPD) who were treated with valproate for 4 weeks exhibited a decrease in histone 3 and 4 acetylation in peripheral blood mononuclear cells [49]. This data was then further supported by a study that found that the ability of chromatin to "open" in those same cells is impaired in SCZ [50], which would suggest that histone modification pathways are dysregulated in SCZ, leading to abnormal chromatin organization and therefore dysregulating gene expression. One small study ($n = 8$) found changes in histone trimethylation in olfactory cells surrounding 22 genes, 3 of which were validated by qPCR [51]; one gene in particular, DAAM2, has even been suggested as a possible biomarker for diagnosis and disease state [52]. Histone modifications have even been seen to occur in the GABAergic system in response to clozapine, an antipsychotic medication commonly used in SCZ [53]. This, along with the fact that other drugs can modify histone PTM profiles, has led to the suggestion of new therapeutic targets in the context of SCZ [54].

As compiled by Akbarian, other potential biomarkers have been found in lymphocyte extracts of schizophrenia patients, though these studies did not combine proteomic identification of histone modifications with sequencing technology, reducing the robustness of the potential biomarkers [55]. Since then, several other histone sites of acetylation, methylation, and phosphorylation [56] have been associated with SCZ in peripheral and central nervous system tissues, supporting the neurodevelopmental hypothesis of schizophrenia and offering other potential biomarkers. Even more recently, less commonly studied modifications were implicated in the pathophysiology of SCZ, including ubiquitination, UFMylation, SUMOylation, and NEDD8ylation, and have been proposed as potential therapeutic targets [57].

A link between bipolar disorder (BD) and histone modification dysregulation has also been studied, and a global increase in histone H3 phosphorylation was found as well as H3 acetylation at a gene of interest (drebrin-like protein) in a small pilot study [3], and other genes of interest were found to be affected differently by aging in patients with BD compared to patients with SCZ [16]. One review compiled a list of HDAC and sirtuin proteins that have been found dysregulated in BD and major depressive disorder (MDD), also covering histone acetylation and trimethylation [58]. These histone modifications in BD, in addition to a large list of other epigenetic effects, have been compiled and covered in more detail in a recent review [59].

7.7 Depression and Chronic Stress

Data specific to MDD are more extensive, including sources that used animal models and postmortem brain tissue. MDD is a disorder that is well known to have both genetic and environmental factors and, as reviewed in detail by Deussing and Jakovcevski, the potentially long-lasting effects of one's upbringing and lifestyle, particularly stress, on epigenetics offers a strong hypothesis to link these two components [60]. Chronic stress is thought to affect epigenetics in multiple ways, including modulating histone acetylation, which has been shown to be associated with fear conditioning, synaptic plasticity, and long-term memory formation [61], though mostly in animal models. Methylation has also been implicated in MDD due to correlations with depression-like behavior and symptoms and stress reactivity [60, 62]. Leaving the realm of correlation and entering causation, a loss of an X-linked demethylase *Phf8* has been shown to lead to a anxiety- and depression-resistant phenotype in mice, through a biological association with serotonergic pathways [63]. These data offer possible therapeutic targets with acetylation- and methylation-based treatment options. In fact, as reviewed by Saavedra et al., histone deacetylase inhibitors (HDACi) have been tested as an adjunctive therapy in several animal models and several current options are hypothesized to have affect histone acetylation [64].

7.8 Concluding Remarks

Strong links between histone modifications and neurodevelopmental and psychiatric disorders have been made; although there is much that has yet to be discovered. The vast majority of studies that have been carried out until only very recently have involved only a few post-translational modifications, focusing mainly on histone acetylation, methylation, and phosphorylation. With advances in proteomic and other 'omic techniques, it has become possible to increase sample number and detect even more modifications at even a sequence level, rather than looking at only overall modification levels. This data has great potential, and has indeed shown some of this potential, to further the current knowledge about how these disorders occur. In the upcoming years, much more comprehensive data regarding the 30+ known (and still unknown) histone modifications can be expected, offering insight into how to better identify, prevent, and treat human diseases.

References

1. Berger, S. L. (2002). Histone modifications in transcriptional regulation. *Current Opinion in Genetics & Development, 12*, 142–148.
2. Narayan, P. J., Lill, C., Faull, R., Curtis, M. A., & Dragunow, M. (2015). Increased acetyl and total histone levels in post-mortem Alzheimer's disease brain. *Neurobiology of Disease, 74*, 281–294.
3. Rao, J. S., Keleshian, V. L., Klein, S., & Rapoport, S. I. (2012). Epigenetic modifications in frontal cortex from Alzheimer's disease and bipolar disorder patients. *Translational Psychiatry, 2*, e132–e132.
4. Smith, A. R., et al. (2021). The histone modification H3K4me3 is altered at the ANK1 locus in Alzheimer's disease brain. *Future Science OA, 7*, FSO665.
5. Kontopoulos, E., Parvin, J. D., & Feany, M. B. (2006). α-synuclein acts in the nucleus to inhibit histone acetylation and promote neurotoxicity. *Human Molecular Genetics, 15*, 3012–3023.
6. Voutsinas, G. E., et al. (2010). Allelic imbalance of expression and epigenetic regulation within the alpha-synuclein wild-type and p.Ala53Thr alleles in Parkinson disease. *Human Mutation, 31*, 685–691.
7. Nicholas, A. P., et al. (2008). Striatal histone modifications in models of levodopa-induced dyskinesia. *Journal of Neurochemistry, 106*, 486–494.
8. Mazzocchi, M., Collins, L. M., Sullivan, A. M., & O'Keeffe, G. W. (2020). The class II histone deacetylases as therapeutic targets for Parkinson's disease. *Neuronal Signaling, 4*, NS20200001.
9. Mu, M.-D., et al. (2020). Therapeutic effect of a histone demethylase inhibitor in Parkinson's disease. *Cell Death & Disease, 11*, 1–16.
10. Gapp, K., Woldemichael, B. T., Bohacek, J., & Mansuy, I. M. (2014). Epigenetic regulation in neurodevelopment and neurodegenerative diseases. *Neuroscience, 264*, 99–111.
11. Bassi, S., Tripathi, T., Monziani, A., Di Leva, F., & Biagioli, M. (2017). Epigenetics of Huntington's disease. In R. Delgado-Morales (Ed.), *Neuroepigenomics in aging and disease* (pp. 277–299). Springer. https://doi.org/10.1007/978-3-319-53889-1_15
12. Naia, L., et al. (2017). Histone deacetylase inhibitors protect against pyruvate dehydrogenase dysfunction in Huntington's disease. *The Journal of Neuroscience, 37*, 2776–2794.

13. Suelves, N., Kirkham-McCarthy, L., Lahue, R. S., & Ginés, S. (2017). A selective inhibitor of histone deacetylase 3 prevents cognitive deficits and suppresses striatal CAG repeat expansions in Huntington's disease mice. *Scientific Reports, 7*, 6082.
14. Hecklau, K., et al. (2021). The effects of selective inhibition of histone deacetylase 1 and 3 in Huntington's disease mice. *Frontiers in Molecular Neuroscience, 14*, 17.
15. Akbarian, S., Beeri, M. S., & Haroutunian, V. (2013). Epigenetic determinants of healthy and diseased brain aging and cognition. *JAMA Neurology, 70*, 711–718.
16. Tang, B., Dean, B., & Thomas, E. A. (2011). Disease- and age-related changes in histone acetylation at gene promoters in psychiatric disorders. *Translational Psychiatry, 1*, e64.
17. Castellano, J. F., et al. (2012). Age-related memory impairment is associated with disrupted multivariate epigenetic coordination in the hippocampus. *PLoS One, 7*, e33249.
18. Peleg, S., et al. (2010). Altered histone acetylation is associated with age-dependent memory impairment in mice. *Science, 328*, 753–756.
19. Burns, A. M., & Gräff, J. (2021). Cognitive epigenetic priming: Leveraging histone acetylation for memory amelioration. *Current Opinion in Neurobiology, 67*, 75–84.
20. Fischer, A., Sananbenesi, F., Wang, X., Dobbin, M., & Tsai, L.-H. (2007). Recovery of learning and memory is associated with chromatin remodelling. *Nature, 447*, 178–182.
21. Coneys, R., & Wood, I. C. (2020). Alzheimer's disease: The potential of epigenetic treatments and current clinical candidates. *Neurodegenerative Disease Management, 10*, 543–558.
22. Elia, L. P., Reisine, T., Alijagic, A., & Finkbeiner, S. (2020). Approaches to develop therapeutics to treat frontotemporal dementia. *Neuropharmacology, 166*, 107948.
23. She, A., et al. (2017). Selectivity and Kinetic requirements of HDAC Inhibitors as progranulin enhancers for treating frontotemporal dementia. *Cell Chemical Biology, 24*, 892–906.e5.
24. Schmitt, M., & Matthies, H. (1979). Biochemical studies on histones of the central nervous system. III. Incorporation of [14C]-acetate into the histones of different rat brain regions during a learning experiment. *Acta Biologica et Medica Germanica, 38*, 683–689.
25. Gupta, S., et al. (2010). Histone methylation regulates memory formation. *The Journal of Neuroscience, 30*, 3589–3599.
26. Kim, S., & Kaang, B.-K. (2017). Epigenetic regulation and chromatin remodeling in learning and memory. *Experimental & Molecular Medicine, 49*, e281.
27. Levenson, J. M., & Sweatt, J. D. (2005). Epigenetic mechanisms in memory formation. *Nature Reviews. Neuroscience, 6*, 108–118.
28. Fischer, A. (2019). The role of dynamic histone modifications in learning behavior. In E. B. Binder & T. Klengel (Eds.), *Behavioral neurogenomics* (pp. 127–157). Springer. https://doi.org/10.1007/7854_2019_108
29. Schmauss, C. (2017). The roles of class I histone deacetylases (HDACs) in memory, learning, and executive cognitive functions: A review. *Neuroscience & Biobehavioral Reviews, 83*, 63–71.
30. Volmar, C.-H., & Wahlestedt, C. (2015). Histone deacetylases (HDACs) and brain function. *Neuroepigenetics, 1*, 20–27.
31. Tapias, A., & Wang, Z.-Q. (2017). Lysine acetylation and deacetylation in brain development and neuropathies. *Genomics, Proteomics & Bioinformatics, 15*, 19–36.
32. Akbarian, S., & Huang, H.-S. (2009). Epigenetic regulation in human brain—Focus on histone lysine methylation. *Biological Psychiatry, 65*, 198–203.
33. Ginsberg, M. R., Rubin, R. A., Falcone, T., Ting, A. H., & Natowicz, M. R. (2012). Brain transcriptional and epigenetic associations with autism. *PLoS One, 7*, e44736.
34. Eshraghi, A. A., et al. (2018). Epigenetics and autism spectrum disorder: Is there a correlation? *Frontiers in Cellular Neuroscience, 12*, 78.
35. Zhang, Q., et al. (2013). Histone modification mapping in human brain reveals aberrant expression of histone H3 lysine 79 dimethylation in neural tube defects. *Neurobiology of Disease, 54*, 404–413.

36. Kleine-Kohlbrecher, D., et al. (2010). A functional link between the histone demethylase PHF8 and the transcription factor ZNF711 in X-linked mental retardation. *Molecular Cell, 38*, 165–178.
37. Burgold, T., et al. (2008). The histone H3 lysine 27-specific demethylase Jmjd3 is required for neural commitment. *PLoS One, 3*, e3034.
38. Lee, E. R., Murdoch, F. E., & Fritsch, M. K. (2007). High histone acetylation and decreased polycomb repressive complex 2 member levels regulate gene specific transcriptional changes during early embryonic stem cell differentiation induced by retinoic acid. *Stem Cells, 25*, 2191–2199.
39. Kuroki, S., et al. (2013). Epigenetic regulation of mouse sex determination by the histone demethylase Jmjd1a. *Science, 341*, 1106–1109.
40. Qureshi, I. A., & Mehler, M. F. (2010). Chapter 6 – Genetic and epigenetic underpinnings of sex differences in the brain and in neurological and psychiatric disease susceptibility. In I. Savic (Ed.), *Progress in brain research* (Vol. 186, pp. 77–95). Elsevier.
41. Ratnu, V. S., Emami, M. R., & Bredy, T. W. (2017). Genetic and epigenetic factors underlying sex differences in the regulation of gene expression in the brain. *Journal of Neuroscience Research, 95*, 301–310.
42. Tsai, H.-W., Grant, P. A., & Rissman, E. F. (2009). Sex differences in histone modifications in the neonatal mouse brain. *Epigenetics, 4*, 47–53.
43. Matsuda, K. I., et al. (2011). Histone deacetylation during brain development is essential for permanent masculinization of sexual behavior. *Endocrinology, 152*, 2760–2767.
44. Xu, J., Deng, X., & Disteche, C. M. (2008). Sex-specific expression of the X-linked histone demethylase gene Jarid1c in brain. *PLoS One, 3*, e2553.
45. Martins-de-Souza, D., et al. (2010). Sex-specific proteome differences in the anterior cingulate cortex of schizophrenia. *Journal of Psychiatric Research, 44*, 989–991.
46. Singh, G., Singh, V., & Schneider, J. S. (2019). Post-translational histone modifications and their interaction with sex influence normal brain development and elaboration of neuropsychiatric disorders. *Biochimica et Biophysica Acta (BBA) - Molecular Basis of Disease, 1865*, 1968–1981.
47. Girdhar, K., et al. (2018). Cell-specific histone modification maps in the human frontal lobe link schizophrenia risk to the neuronal epigenome. *Nature Neuroscience, 21*, 1126–1136.
48. Curtis, D. (2016). Pathway analysis of whole exome sequence data provides further support or the involvement of histone modification in the aetiology of schizophrenia. *Psychiatric Genetics, 26*, 223–227.
49. Sharma, R. P., et al. (2006). Valproic acid and chromatin remodeling in schizophrenia and bipolar disorder: Preliminary results from a clinical population. *Schizophrenia Research, 88*, 227–231.
50. Gavin, D. P., & Sharma, R. P. (2010). Histone modifications, DNA methylation, and schizophrenia. *Neuroscience & Biobehavioral Reviews, 34*, 882–888.
51. Kano, S., et al. (2013). Genome-wide profiling of multiple histone methylations in olfactory cells: Further implications for cellular susceptibility to oxidative stress in schizophrenia. *Molecular Psychiatry, 18*, 740–742.
52. Kuzman, M. R., Medved, V., Terzic, J., & Krainc, D. (2009). Genome-wide expression analysis of peripheral blood identifies candidate biomarkers for schizophrenia. *Journal of Psychiatric Research, 43*, 1073–1077.
53. Huang, H.-S., et al. (2007). Prefrontal dysfunction in schizophrenia involves mixed-lineage leukemia 1-regulated histone methylation at GABAergic gene promoters. *The Journal of Neuroscience, 27*, 11254–11262.
54. Peter, C. J., & Akbarian, S. (2011). Balancing histone methylation activities in psychiatric disorders. *Trends in Molecular Medicine, 17*, 372–379.
55. Akbarian, S. (2014). Epigenetic mechanisms in schizophrenia. *Dialogues in Clinical Neuroscience, 16*, 405–417.

56. Thomas, E. A. (2017). Histone posttranslational modifications in schizophrenia. In R. Delgado-Morales (Ed.), *Neuroepigenomics in Aging and Disease* (pp. 237–254). Springer. https://doi.org/10.1007/978-3-319-53889-1_13
57. Abdolmaleky, H. M., & Thiagalingam, S. (2021). Chapter 14 – Pathogenic histone modifications in schizophrenia are targets for therapy. In J. Peedicayil, D. R. Grayson, & D. Avramopoulos (Eds.), *Epigenetics in psychiatry* (2nd ed., pp. 309–319). Academic Press. https://doi.org/10.1016/B978-0-12-823577-5.00006-4
58. Ludwig, B., & Dwivedi, Y. (2016). Dissecting bipolar disorder complexity through epigenomic approach. *Molecular Psychiatry, 21*, 1490–1498.
59. Legrand, A., Iftimovici, A., Khayachi, A., & Chaumette, B. (2021). Epigenetics in bipolar disorder: A critical review of the literature. *Psychiatric Genetics, 31*, 1–12.
60. Deussing, J. M., & Jakovcevski, M. (2017). Histone modifications in major depressive disorder and related rodent models. In R. Delgado-Morales (Ed.), *Neuroepigenomics in aging and disease* (pp. 169–183). Springer. https://doi.org/10.1007/978-3-319-53889-1_9
61. Nagy, C., Vaillancourt, K., & Turecki, G. (2018). A role for activity-dependent epigenetics in the development and treatment of major depressive disorder. *Genes, Brain and Behavior, 17*, e12446.
62. Wan, Q., et al. (2014). Histone modifications of the Crhr1 gene in a rat model of depression following chronic stress. *Behavioural Brain Research, 271*, 1–6.
63. Walsh, R. M., et al. (2017). Phf8 loss confers resistance to depression-like and anxiety-like behaviors in mice. *Nature Communications, 8*, 15142.
64. Saavedra, K., Molina-Márquez, A. M., Saavedra, N., Zambrano, T., & Salazar, L. A. (2016). Epigenetic modifications of major depressive disorder. *International Journal of Molecular Sciences, 17*, 1279.
65. Arnaudo, A. M., & Garcia, B. A. (2013). Proteomic characterization of novel histone post-translational modifications. *Epigenetics & Chromatin, 6*(1), 10.1186/1756-8935-6-24.
66. Bao, X., Liu, Z., Zhang, W., Gladysz, K., Fung, Y. M. E., Tian, G., Xiong, Y., Wong, J. W. H., Yuen, K. W. Y., & Li, X. D. (2019). Glutarylation of histone H4 lysine 91 regulates chromatin dynamics. *Molecular Cell, 76*(4), 660–675.e9. https://doi.org/10.1016/j.molcel.2019.08.018
67. Berger, S. L. (2002). Histone modifications in transcriptional regulation. *Current Opinion in Genetics & Development, 12*(2), 142–148. https://doi.org/10.1016/S0959-437X(02)00279-4
68. Brown, J. S., & Jackson, S. P. (2015). Ubiquitylation neddylation and the DNA damage response. *Open Biology, 5*(4), 150018. https://doi.org/10.1098/rsob.150018
69. Chen, Y., Sprung, R., Tang, Y., Ball, H., Sangras, B., Kim, S. C., Falck, J. R., Peng, J., Wei, G., & Zhao, Y. (2007). Lysine Propionylation and Butyrylation are novel post-translational modifications in histones. *Molecular & Cellular Proteomics, 6*(5), 812–819. https://doi.org/10.1074/mcp.M700021-MCP200
70. Dai, L., Peng, C., Montellier, E., Zhike, L., Chen, Y., Ishii, H., Debernardi, A., Buchou, T., Rousseaux, S., Jin, F., Sabari, B. R., Zhiyou Deng, C., Allis, D., Ren, B., Khochbin, S., & Zhao, Y. (2014). Lysine 2-hydroxyisobutyrylation is a widely distributed active histone mark. *Nature Chemical Biology, 10*(5), 365–370. https://doi.org/10.1038/nchembio.1497
71. García-Giménez, J. L., Olaso, G., Hake, S. B., Bönisch, C., Wiedemann, S. M., Markovic, J., Dasí, F., Gimeno, A., Pérez-Quilis, C., Palacios, Ò., Capdevila, M., Viña, J., & Pallardó, F. V. (2013). Histone H3 glutathionylation in proliferating mammalian cells destabilizes nucleosomal structure. *Antioxidants & Redox Signaling, 19*(12), 1305–1320. https://doi.org/10.1089/ars.2012.5021
72. García-Giménez, J. L., Ledesma, A. M. V., Esmoris, I., Romá-Mateo, C., Sanz, P., Viña, J., & Pallardó, F. V. (2012). Histone carbonylation occurs in proliferating cells. *Free Radical Biology and Medicine, 52*(8), 1453–1464. https://doi.org/10.1016/j.freeradbiomed.2012.01.022
73. Hassan, Y. I., & Zempleni, J. A novel enigmatic histone modification: Biotinylation of histones by holocarboxylase synthetase. *Nutrition Reviews, 66*(12), 721–725. https://doi.org/10.1111/j.1753-4887.2008.00127.x

74. He, H., Zhang, D., Wang, Y., Perez-Neut, M., Zhen, H., George Zheng, Y., Hao, Q., & Zhao, Y. (2018). Lysine benzoylation is a histone mark regulated by SIRT2. *Nature Communications, 9*(1). https://doi.org/10.1038/s41467-018-05567-w
75. Jiang, T., Zhou, X., Taghizadeh, K., Dong, M., & Dedon, P. C. (2007). N-formylation of lysine in histone proteins as a secondary modification arising from oxidative DNA damage. *Proceedings of the National Academy of Sciences, 104*(1), 60–65. https://doi.org/10.1073/pnas.0606775103
76. Kebede, A. F., Nieborak, A., Shahidian, L. Z., Le, S., Richter, G. F., Gómez, D. A., Baltissen, M. P., Meszaros, G., de Fatima, H., Magliarelli, A. T., Margueron, R., Colomé-Tatché, M., Ricci, R., Daujat, S., Vermeulen, M., Mittler, G., & Schneider, R. (2017). Histone propionylation is a mark of active chromatin. *Nature Structural & Molecular Biology, 24*(12), 1048–1056. https://doi.org/10.1038/nsmb.3490
77. Kothapalli, N., Camporeale, G., Kueh, A., Chew, Y. C., Oommen, A. M., Griffin, J. B., & Zempleni, J. (2005). Biological functions of biotinylated histones. *The Journal of Nutritional Biochemistry, 16*(7), 446–448. https://doi.org/10.1016/j.jnutbio.2005.03.025
78. Kouzarides, T. (2007). Chromatin modifications and their function. *Cell, 128*(4), 693–705. https://doi.org/10.1016/j.cell.2007.02.005
79. Kwok, J., O'Shea, M., Hume, D. A., & Lengeling, A. (2017). Jmjd6 a JmjC dioxygenase with many interaction partners and pleiotropic functions. *Frontiers in Genetics, 8*. https://doi.org/10.3389/fgene.2017.00032
80. Liu, S., Liu, G., Cheng, P., Xue, C., Zhou, Y., Xu, C., Lu, Y., Qiao, Z., Zhang, T., & Gong, Z. (2019). Genome-wide profiling of histone lysine Butyrylation reveals its role in the positive regulation of gene transcription in Rice. *Rice, 12*(1). https://doi.org/10.1186/s12284-019-0342-6
81. Qin, B., Yu, J., Nowsheen, S., Wang, M., Xinyi, T., Liu, T., Li, H., Wang, L., & Lou, Z. (2019). UFL1 promotes histone H4 ufmylation and ATM activation. *Nature Communications, 10*(1). https://doi.org/10.1038/s41467-019-09175-0
82. Tan, M., Luo, H., Lee, S., Jin, F., Yang, J. S., Montellier, E., Buchou, T., Cheng, Z., Rousseaux, S., Rajagopal, N., Zhike, L., Ye, Z., Zhu, Q., Wysocka, J., Ye, Y., Khochbin, S., Ren, B., & Zhao, Y. (2011). Identification of 67 histone Marks and histone lysine crotonylation as a new type of histone modification. *Cell, 146*(6), 1016–1028. https://doi.org/10.1016/j.cell.2011.08.008
83. Unoki, M., Masuda, A., Dohmae, N., Arita, K., Yoshimatsu, M., Iwai, Y., Fukui, Y., Ueda, K., Hamamoto, R., Shirakawa, M., Sasaki, H., & Nakamura, Y. (2013). Lysyl 5-hydroxylation a novel histone modification by Jumonji domain containing 6 (JMJD6)*. *Journal of Biological Chemistry, 288*(9), 6053–6062. https://doi.org/10.1074/jbc.M112.433284
84. Wilson, J. P., Raghavan, A. S., Yang, Y.-Y., Charron, G., & Hang, H. C. (2011). Proteomic analysis of fatty-acylated proteins in mammalian cells with chemical reporters reveals S-acylation of histone H3 variants. *Molecular & Cellular Proteomics, 10*(3), M110.001198. https://doi.org/10.1074/mcp.M110.001198
85. Wondrak, G. T., Cervantes-Laurean, D., Jacobson, E. L., & Jacobson, M. K. (2000). Histone carbonylation in vivo and in vitro. *Biochemical Journal, 351*(3), 769–777. https://doi.org/10.1042/bj3510769
86. Xie, Z., Dai, J., Dai, L., Tan, M., Cheng, Z., Yeming, W., Boeke, J. D., & Zhao, Y. (2012). Lysine Succinylation and Lysine Malonylation in Histones. *Molecular & Cellular Proteomics, 11*(5), 100–107. https://doi.org/10.1074/mcp.M111.015875
87. Zhang, D., Tang, Z., Huang, H., Zhou, G., Cui, C., Weng, Y., Liu, W., Kim, S., Lee, S., Perez-Neut, M., Ding, J., Czyz, D., Rong, H., Ye, Z., He, M., Zheng, Y. G., Shuman, H. A., Dai, L., Ren, B., … Zhao, Y. (2019). Metabolic regulation of gene expression by histone lactylation. *Nature, 574*(7779), 575–580. https://doi.org/10.1038/s41586-019-1678-1
88. Zhao, S., Yue, Y., Li, Y., & Li, H. (2019). Identification and characterization of 'readers' for novel histone modifications. *Current Opinion in Chemical Biology, 5157-65*. https://doi.org/10.1016/j.cbpa.2019.04.001

89. Zhao, S., Zhang, X., & Li, H. (2018). Beyond histone acetylation—Writing and erasing histone acylations. *Current Opinion in Structural Biology, 53*, 169–177. https://doi.org/10.1016/j.sbi.2018.10.001
90. Zhu, Z., Han, Z., Halabelian, L., Yang, X., Ding, J., Zhang, N., Ngo, L., Song, J., Zeng, H., He, M., Zhao, Y., Arrowsmith, C. H., Luo, M., Bartlett, M. G., & George Zheng, Y. (2021). (2020) Identification of lysine isobutyrylation as a new histone modification mark. *Nucleic Acids Research, 49*(1), 177–189. https://doi.org/10.1093/nar/gkaa1176

Chapter 8
Mitochondrial Dysregulation and the Influence in Neurodegenerative Diseases

Giuliana S. Zuccoli and Victor Corasolla Carregari

Abstract Mitochondrial function is essential to ensure vital cellular processes. Given the energy requirement of the brain, neuronal function, viability, and survival are closely related to proper mitochondrial function. Dysregulation of mitochondrial processes can lead to several detrimental effects in the cells and stablish the condition of mitochondrial dysfunction. This dysfunction is proposed to be greatly implicated in several neurodegenerative diseases, with evidence of compromised mitochondrial function and dynamics in Alzheimer's disease, Parkinson's disease, Huntington's disease, and amyotrophic lateral sclerosis.

Keywords Neurogenerative diseases · Mitochondrial dysfunction · Proteomics · Brain metabolism

8.1 Introduction

Mitochondria are intracellular organelles that play a part in key cellular processes, such as ATP production, calcium buffering, redox homeostasis, and cell death control [1, 2]. Given the high energy demand of the brain, the role of the organelle becomes especially important for the proper function of this organ [3]. In addition to the overall roles attributed to mitochondria, in neurons, these organelles are also crucial for neurotransmission, plasticity and membrane excitability [4]. Mitochondria are dynamic organelles, with constant changes in size, shape, permeability and number depending on cellular function [5]. Interestingly, unlike other organelles, mitochondria present their own DNA (mtDNA) and are able to transcribe and translate it independent from the nuclear DNA, even though mitochondrial and cellular function are interdependent [6].

G. S. Zuccoli (✉) · V. C. Carregari
Laboratory of Neuroproteomics, Department of Biochemistry and Tissue Biology, Institute of Biology, University of Campinas (UNICAMP), Campinas, SP, Brazil

V. Corasolla Carregari (ed.), *Understanding PTMs in Neurodegenerative Diseases*, Advances in Experimental Medicine and Biology 1382,
https://doi.org/10.1007/978-3-031-05460-0_8

The dynamic and complex function of these organelles end up making them more prone to damage throughout the lifespan of the cell. The accumulated damage can in turn lead to mitochondrial dysfunction, characterized as decreased ATP production or decreased electron transport chain efficiency, reactive oxygen species (ROS) unbalance potentially leading to oxidative stress, and disturbed calcium buffering. Mitochondrial dysfunction, caused by either aging or chronic illness may cause progression of cell death leading to neuronal loss [4, 7–10]. In fact, is proposed that mitochondrial defects contribute to neuronal degeneration in age-related neurodegenerative disorders, such as Alzheimer's disease, Parkinson's disease, Huntington's disease and amyotrophic lateral sclerosis [11–14].

8.2 Neurodegenerative Diseases and Evidence of Mitochondrial Dysfunction

Neurodegenerative diseases are a heterogeneous group of debilitating disorders that are characterized by the progressive and selective loss of neuronal systems [15, 16]. Even though those disorders are usually defined by specific protein depositions and anatomic features, they share several processes related to neuronal dysfunction and subsequent neuronal death [17]. Despite of the specific mechanisms of each disease that lead to neurodegeneration, they are known to share some common features, such as mitochondrial dysfunction [16, 18]. Among the neurodegenerative disease, in here we will discuss the evidence of mitochondrial dysfunction in Alzheimer's disease, Parkinson's disease, Huntington's disease and amyotrophic lateral sclerosis.

8.2.1 Parkinson's Disease

Parkinson's disease is one of the most prevalent neurodegenerative diseases, with an estimation of 0.3% prevalence worldwide, which dramatically increases to over 3% in individuals with 80 years of age or more [19]. Before 50 years of age, the disease onset is rare, but from the sixth decade of life forward the chances of developing Parkinson's increase up to tenfold [20–22]. The clinical diagnosis of the disease arises from the presence of bradykinesia, which is characterized as slowness of movement, along with rigidity and/or rest tremor. An important part of the diagnosis consists of determining that the cause of Parkinsonism is Parkinson's disease, which can include clinical or imaging evidence, supportive criteria such as L-DOPA responsiveness, among others [23–26]. The average age of onset in on the late 50s and young onset is characterized by an age of less than 45 years of age. The earlier the onset, there is an increased proportion of the disease being genetically defined [27, 28]. Non-motor symptoms are also common in the progression of the disease and can precede de motor symptoms by years [29] and are characterized as pain, mood disorders, cognitive impairment, sleep disorders and others [30].

The combination of neuronal loss, predominantly dopaminergic, in areas of the substantia nigra and widespread intracellular α-synuclein accumulation is an important characteristic of Parkinson disease [29]. Several theories have been proposed to try to elucidate the pathogenesis of Parkinson's disease, with an important role of mitochondrial dysfunction in both familial and sporadic forms of the illness [31, 32]. This dysfunction can be attributed to either environmental factor or genetic mutations [33]. The first study that demonstrated the involvement of mitochondrial respiration in the disease pathogenesis consisted of the observation that self-injected infusions of the toxin 1-methyl-4-phenyl-1, 2, 3, 6-tetrahydrodropyridine (MPTP), a selective inhibitor of the mitochondrial complex I (MCI) of the electron transport chain (ETC), lead to the appearance of parkinsonism [34]. Impairments in MCI activity have been observed in the substantia nigra, skeletal muscles and platelets of Parkinson's disease patients [15, 32, 35]. In line with that, rotenone, which is a MCI inhibitor, is administered in rodents to create a mouse model of Parkinson's disease and in a recent study that conducted genetic manipulation causing the disruption of the MCI in dopaminergic neurons of mice observed that MCI dysfunction culminated in the loss of the dopaminergic phenotype accompanied by motor deficits [34].

Mitochondrial DNA (mtDNA) mutations have been implicated in the neuronal degeneration characteristic of the disease, as it was observed that after transfection of patients' mtDNA into normal cells resulted in deficits in MCI activity, lower mitochondrial membrane potential and altered calcium homeostasis in recipient cells [4, 36]. Higher levels of mtDNA mutations have been observed in neurons from the substantia nigra of patients and different genes involved with the occurrence of Parkinson disease are associated with mitochondrial function [37, 38]. Proteins associated with the disease, such as leucine-reach repeat kinase 2 (LRRK2), α-synuclein and protein deglycase (DJ-1), are involved in mitochondrial function in the disease and are involved in hereditary and potentially also sporadic cases of the disease [39]. LRRK2 interacts with regulators of the fission and fusion mitochondrial machinery, including dynamin-related protein 1 (DRP1), and therefore, is the regulation of this machinery is proposed to be an important feature of the disease [39, 40]. Overexpression of α-synuclein was shown to impact mitochondrial function and there have been reports of physical association between the protein and DJ-1, which in turn act as a protective mechanism against cell death induced by oxidative stress [18, 33]. There is a link between DJ-1 mutation mitochondrial alterations, such as impaired respiration, increased ROS levels and altered morphology [41]. Mutations in PTEN-induced kinase 1 (PINK1) and Parkin have been associated with early-onset Parkinson's disease [36] and there is also a link with neurotoxicity and mitochondrial dysfunction. The loss or improper function of Parkin can have an effect on the release of cytochrome c and mitochondrial biogenesis [14, 42]. Therefore, there is an association between the deletion of the mitochondrial proteins PINK1 and Parkin with the neurodegeneration observed in the disease [10], given that PINK1 presents a protective effect against cell death [14, 42]. Also, PINK1 mutations in cell culture models are involved with decreased respiration and ATP synthesis, as well as aggregation of α-synuclein [43].

8.2.2 Alzheimer's Disease

Alzheimer's disease, which is an age-related neurodegenerative disease, is the most prevalent form of dementia and the most common neurodegenerative disorder, with a high incidence in people over the age of 65 [44]. People with the disorder present memory loss and compromised cognitive function, which is explained by the severe neuronal loss in the brain [45, 46]. The presence of β-amyloid (Aβ)-containing extracellular plaques and tau-containing intracellular neurofibrillary tangles biologically define the disease [47]. These Aβ plaques arise from the improper processing of the amyloid-precursor protein (APP) in specific brain areas. Most of the cases of Alzheimer's disease are sporadic, with 10% of cases caused by genetic mutations in the genes of APP, presenilin 1 and presenilin 2, which are involved in Aβ production [48].

Mitochondrial dysfunction is proposed to underly the neurodegeneration seen in the disease, as evidence of altered mitochondrial morphology, abnormal energy metabolism in the brain, elevated ROS levels, disruption of calcium homeostasis and mtDNA alterations arise in the context of the disorder [18, 49]. In fact, some studies show that impairments in brain metabolism are present decades before the appearance of dementia [50–52]. There is evidence of accumulation of APP in the mitochondrial channels, leading to mitochondrial toxicity and cell damage [53]. This accumulation affects mitochondrial respiration, with dysregulation of complex I and IV of the electron transport chain [54, 55]. One proteomic study revealed the role for the 42 amino acids long beta-amyloid (Aβ42) peptide in disrupting proteome responses for signaling, bioenergetics, and morphology in mitochondria [56] and elevated levels of this fragment are believed to contribute to neural damage in both early and late stages of Alzheimer's disease [57]. Phosphorylated tau also seems to have a connection with mitochondrial dysfunction by blocking mitochondrial transport and leading to energetic deficits and increased ROS levels at the synapse [53, 58].

The presence of oxidative damage is an important feature of Alzheimer's disease, even before the appearance of significant plaque accumulation [59] and it has been proposed that oxidative stress may actually have an effect on signaling pathways that are involved with APP and tau processing [15]. In turn, it has been observed the interaction between Aβ and the mitochondrial voltage-dependent anion channel 1 (VDAC1), which leads to impaired protein transport and increased ROS production [49, 60]. Another important observation in the context of mitochondrial function in Alzheimer's disease is the pronounced alteration in mitochondrial dynamics, characterized by increased mitochondrial fission and changes in mtDNA [61]. Mitochondrial fission depends on the dynamin-related protein 1 (DRP1) and with was discovered that the interaction between DRP1 and Aβ leads to reduced fusion and increased mitochondrial fragmentation [62]. Along with Aβ, tau is also known to affect mitochondrial fission by interacting with important regulators of this process [63]. Impairments of mitochondrial axonal transport are linked to disturbed synaptic function in the disease and tau was shown to hinder mitochondrial transport to the synapse, affecting neurotransmitter release and overall synaptic function [64].

8.2.3 *Huntington's Disease*

Huntington's disease is a monogenic neurodegenerative disorder that has an autosomal dominant inheritance. It presents a combination of cognitive, motor, and behavioral symptoms and is caused by the mutant form of the protein huntingtin, which presents toxic features to neurons, leading to their dysfunction and death [65, 66]. The mutant form of the protein occurs due to a polymorphic CAG trinucleotide repeat in the coding sequence of huntingtin, resulting in an expanded polyglutamine repeat in the protein that have profound effects on the striatum and cerebral cortex [65].

Many studies have found a correlation between mitochondrial dysfunction and the pathogenesis of Huntington's disease, such as decreased activities of complex II and III of the ETC in the brains of patients with the disease [15, 67]. Elevated levels and accumulation of ROS, leading to the establishment of oxidative stress, are also believed to be important features of the disorder [14]. It is proposed that the generation of ROS interact with and activate proteins involved with mitochondrial dynamics, such as mitochondrial fission 1 (FIS1), causing increased mitochondrial fragmentation that could lead to neuronal damage [68]. The mutated form of huntingtin interacts with DRP1, enhancing its activity, which leads to increased fragmentation and disturbed mitochondrial distribution. This in turn can promote malfunctioning mitochondrial transport along axon and cause synaptic degeneration [14].

There is evidence of the repression of peroxisome proliferator-activated receptor gamma coactivator-1α (PGC-1α) in Huntington's disease, due to interaction with the mutated form of huntingtin and direct promoter binding [69]. PGC-1α is a transcription coactivator that acts on important mitochondrial processes, such as oxidative phosphorylation and mitochondrial biogenesis. It also acts on regulating the transcription of subunits of complexes of the ETC, ATP synthase, superoxide dismutase 2 (SOD2) and mitochondrial transcription factor A (TFAM) [70, 71]. Given the role of PGC-1α and evidence of its compromised function in Huntington's disease, it is proposed that this deficiency can be linked to the mitochondrial dysfunction and oxidative damage observed in the disease [72, 73].

8.2.4 *Amyotrophic Lateral Sclerosis*

Amyotrophic lateral sclerosis is a heterogenous neurodegenerative disorder characterized by the degeneration of upper and lower motor neurons, first leading to muscle weakness and then paralysis [74]. Some individuals present a familial disease, but for most patients, the causes of the disorder are unknown [75]. Even being a motor neuron disorder, approximately 50% of patients present behavioral and/or cognitive impairments, and up to 13% of patients present frontotemporal dementia [76–78]. Thus, characterizing the disease as neurodegenerative rather than neuromuscular. The disease is characterized by the presence of cytosolic protein aggregates, which affect cellular processes, leading to oxidative stress, mitochondrial dysfunction, axonal and synaptic deficits, and cell death [79, 80].

Exacerbated production and accumulation of ROS, leading to the establishment of oxidative stress, is proposed to be an important contributor for the disease's pathology. Familial cases of the disorder account for 10% of the total number of cases and 20% of the familial cases are caused by the expression of a mutated form of the Cu/Zn-superoxide dismutase (SOD1) [15, 79]. The mutated SOD1 interacts with apoptotic regulators, interferes with VDAC function, promotes calcium accumulation, and decreases ATP production, which affects mitochondrial function and leads to increased productions of ROS [14]. Besides, mutated SOD1 reduces the activity of complexes II and IV of the ETC [81]. Other mitochondrial-related mutations are also linked to the disease, such as fused in sarcoma/translocated in liposarcoma (FUS), TAR DNA-binding protein 43 (TDP- 43), and C9ORF72 [82]. Both FUS and TDP-43 are reported to be recruited in mitochondria and mutated forms of those proteins are known to interfere with mitochondrial dynamics, mitochondrial degradation, calcium homeostasis, endoplasmic reticulum-mitochondria association, mitochondrial axonal transport and ROS generation [81]. Altogether, those mitochondrial alterations observed in the context of amyotrophic lateral sclerosis could help explain the presynaptic decline in neuromuscular junctions and subsequent neurodegeneration observed in the disease [14, 80].

8.3 Concluding Remarks

Mitochondria are organelles involved in many vital cellular processes, such as energy metabolism, calcium homeostasis, ROS production and apoptosis. Mitochondrial function in the context of the brain ensures synaptic transmission, neural cell homeostasis and consequently, proper brain function. Mitochondrial dysfunction and oxidative stress are early features of neurodegenerative disorders and are even proposed to have a causal role in the development of those diseases. Evidence of electron transport chain deficits, elevated ROS levels, altered mitochondrial dynamics, mtDNA mutations and detrimental association of disease-specific related proteins with mitochondria are commonly observed in neurodegenerative diseases, such as Alzheimer's disease, Parkinson's disease, Huntington's disease, and amyotrophic lateral sclerosis. For that reason, studying mitochondrial function in the context of neurodegenerative diseases is important for a better understanding of the establishment of these disorders and could provide promising new targets for therapeutic interventions.

References

David, G., & Barett, E. F. (2003). Mitochondrial Ca2+ uptake prevents desynchronization of quantal release and minimizes depletion during repetitive stimulation of mouse motor nerve terminals. *The Journal of Physiology, 548*, 425–438.

Jonckheere, A. I., Smeitink, J. A. M., & Rodenburg, R. J. T. (2012). Mitochondrial ATP synthase: Architecture, function and pathology. *Journal of Inherited Metabolic Disease, 35*, 211–225.

Magistretti, P. J. (2004). Brain energy metabolism. In *From molecules to networks* (pp. 67–89). Elsevier. https://doi.org/10.1016/B978-012148660-0/50004-4

Norat, P., et al. (2020). Mitochondrial dysfunction in neurological disorders: Exploring mitochondrial transplantation. *npj Regenerative Medicine, 5*, 1–9.

Picard, M., Taivassalo, T., Gouspillou, G., & Hepple, R. T. (2011). Mitochondria: Isolation, structure and function. *The Journal of Physiology, 589*, 4413–4421.

Johns, D., & Mitochondrial, R. (1995). DNA and disease. *The New England Journal of Medicine, 333*, 638–644.

Angelova, P. R., & Abramov, A. Y. (2018). Role of mitochondrial ROS in the brain: From physiology to neurodegeneration. *FEBS Letters, 592*, 692–702.

Green, D. R., Galluzzi, L., & Kroemer, G. (2011). Mitochondria and the autophagy–inflammation–cell death axis in organismal aging. *Science (80-.), 333*, 1109–1112.

Park, J. H., & Hayakawa, K. (2021). Extracellular mitochondria signals in CNS disorders. *Frontiers in Cell and Development Biology, 9*, 1–9.

Kumari, R., Shekhar, N., Tyagi, S., & Thakur, A. K. (2021). Mitochondrial dysfunctions and neurodegenerative diseases: A mini-review. https://doi.org/10.15406/japlr.2021.10.00378

Wallace, D. C. (2005). A mitochondrial paradigm of metabolic and degenerative diseases, aging, and cancer: A dawn for evolutionary medicine. *Annual Review of Genetics, 39*, 359–407.

Reddy, P. H., & Beal, M. F. (2005). Are mitochondria critical in the pathogenesis of Alzheimer's disease? *Brain Research Reviews, 49*, 618–632.

Mani, S., et al. (2021). Mitochondrial defects: An emerging theranostic avenue towards Alzheimer's associated dysregulations. *Life Sciences, 285*, 119985.

Elfawy, H. A., & Das, B. (2019). Crosstalk between mitochondrial dysfunction, oxidative stress, and age related neurodegenerative disease: Etiologies and therapeutic strategies. *Life Sciences, 218*, 165–184.

Lin, M. T., & Beal, M. F. (2006). Mitochondrial dysfunction and oxidative stress in neurodegenerative diseases. *Nature, 443*, 787–795.

Gao, J., et al. (2017). Abnormalities of mitochondrial dynamics in neurodegenerative diseases. *Antioxidants, 6*, 25.

Dugger, B. N., & Dickson, D. W. (2017). Pathology of neurodegenerative diseases. *Cold Spring Harbor Perspectives in Biology, 9*, a028035.

Rey, F., Ottolenghi, S., Zuccotti, G., Samaja, M., & Carelli, S. (2022). Mitochondrial dysfunctions in neurodegenerative diseases: Role in disease pathogenesis, strategies for analysis and therapeutic prospects. *Neural Regeneration Research, 17*, 754–758.

Pringsheim, T., Jette, N., Frolkis, A., & Steeves, T. D. L. (2014). The prevalence of Parkinson's disease: A systematic review and meta-analysis. *Movement Disorders, 29*, 1583–1590.

Twelves, D., Perkins, K. S. M., & Counsell, C. (2003). Systematic review of incidence studies of Parkinson's disease. *Movement Disorders, 18*, 19–31.

Savica, R., Grossardt, B. R., Bower, J. H., Ahlskog, J. E., & Rocca, W. A. (2013). Incidence and pathology of synucleinopathies and tauopathies related to Parkinsonism. *JAMA Neurology, 70*, 859–866.

Van Den Eeden, S. K., et al. (2003). Incidence of Parkinson's disease: Variation by age, gender, and race/ethnicity. *American Journal of Epidemiology, 157*, 1015–1022.

Kalia, L. V., & Lang, A. E. (2015). Parkinson's disease. *Lancet, 386*, 896–912.

Tolosa, E., Wenning, G., & Poewe, W. (2006). The diagnosis of Parkinson's disease. *Lancet Neurology, 5*, 75–86.

Gibb, W. R., & Lees, A. J. (1988). The relevance of the Lewy body to the pathogenesis of idiopathic Parkinson's disease. *Journal of Neurology, Neurosurgery, and Psychiatry, 51*, 745–752.
Postuma, R. B., et al. (2015). MDS clinical diagnostic criteria for Parkinson's disease. *Movement Disorders, 30*, 1591–1601.
Alcalay, R. N., et al. (2010). Frequency of known mutations in early onset PD; implication for genetic counseling: The CORE-PD study. *Archives of Neurology, 67*, 1116–1122.
Marder, K. S., et al. (2010). Predictors of Parkin mutations in early-onset Parkinson disease. *Archives of Neurology, 67*, 1–7.
Poewe, W., et al. (2017). Parkinson disease. *Nature Reviews Disease Primers, 3*, 1–21.
Chaudhuri, K. R., & Schapira, A. H. (2009). Non-motor symptoms of Parkinson's disease: Dopaminergic pathophysiology and treatment. *Lancet Neurology, 8*, 464–474.
Schapira, A. H. V. (2007). Mitochondrial dysfunction in Parkinson's disease. *Cell Death and Differentiation, 14*, 1261–1266.
Bose, A., & Beal, M. F. (2016). Mitochondrial dysfunction in Parkinson's disease. *Journal of Neurochemistry, 139*, 216–231.
Monzio Compagnoni, G., et al. (2020). The role of mitochondria in neurodegenerative diseases: The lesson from Alzheimer's disease and Parkinson's disease. *Molecular Neurobiology, 57*, 2959–2980.
González-Rodríguez, P., et al. (2021). Disruption of mitochondrial complex I induces progressive parkinsonism. *Nature, 599*, 650–656.
Beal, M. F. (2005). Mitochondria take center stage in aging and neurodegeneration. *Annals of Neurology, 58*, 495–505.
Gu, M., Cooper, J. M., Taanman, J. W., & Schapira, A. H. V. (1998). Mitochondrial DNA transmission of the mitochondrial defect in Parkinson's disease. *Annals of Neurology, 44*, 177–186.
Johnson, M. E., & Bobrovskaya, L. (2015). An update on the rotenone models of Parkinson's disease: Their ability to reproduce the features of clinical disease and model gene-environment interactions. *Neurotoxicology, 46*, 101–116.
Luo, Y., Hoffer, A., Hoffer, B., & Qi, X. (2015). Mitochondria: A therapeutic target for Parkinson's disease? *International Journal of Molecular Sciences, 16*, 20704–20730.
Ryan, B. J., Hoek, S., Fon, E. A., & Wade-Martins, R. (2015). Mitochondrial dysfunction and mitophagy in Parkinson's: From familial to sporadic disease. *Trends in Biochemical Sciences, 40*, 200–210.
Stafa, K., et al. (2014). Functional interaction of Parkinson's disease-associated LRRK2 with members of the dynamin GTPase superfamily. *Human Molecular Genetics, 23*, 2055–2077.
Krebiehl, G., et al. (2010). Reduced basal autophagy and impaired mitochondrial dynamics due to loss of Parkinson's disease-associated protein DJ-1. *PLoS One, 5*, e9367.
Luoma, P., et al. (2004). Parkinsonism, premature menopause, and mitochondrial DNA polymerase γ mutations: Clinical and molecular genetic study. *Lancet, 364*, 875–882.
Liu, G., et al. (2009). α-Synuclein is differentially expressed in mitochondria from different rat brain regions and dose-dependently down-regulates complex I activity. *Neuroscience Letters, 454*, 187–192.
Knopman, D. S., et al. (2021). Alzheimer disease. *Nature Reviews Disease Primers, 7*, 1–21.
Reddy, P. H., & Beal, M. F. (2008). Amyloid beta, mitochondrial dysfunction and synaptic damage: Implications for cognitive decline in aging and Alzheimer's disease. *Trends in Molecular Medicine, 14*, 45–53.
Reddy, P. H. (2006). Amyloid precursor protein-mediated free radicals and oxidative damage: Implications for the development and progression of Alzheimer's disease. *Journal of Neurochemistry, 96*, 1–13.
Selkoe, D. J. (2001). Alzheimer's disease: Genes, proteins, and therapy. *Physiological Reviews, 81*, 741–766.
Wang, X., et al. (2014). Oxidative stress and mitochondrial dysfunction in Alzheimer's disease. *Biochimica et Biophysica Acta (BBA) - Molecular Basis of Disease, 1842*, 1240–1247.
Shoshan-Barmatz, V., Nahon-Crystal, E., Shteinfer-Kuzmine, A., & Gupta, R. (2018). VDAC1, mitochondrial dysfunction, and Alzheimer's disease. *Pharmacological Research, 131*, 87–101.

Gibson, G. E., & Shi, Q. (2010). A mitocentric view of Alzheimer's disease suggests multi-faceted treatments. *Journal of Alzheimer's Disease, 20*, S591–S607.

Khan, U. A., et al. (2014). Molecular drivers and cortical spread of lateral entorhinal cortex dysfunction in preclinical Alzheimer's disease. *Nature Neuroscience, 17*, 304–311.

Reiman, E. M., et al. (2004). Functional brain abnormalities in young adults at genetic risk for late-onset Alzheimer's dementia. *Proceedings of the National Academy of Sciences of the United States of America, 101*, 284–289.

Reddy, P. H. (2011). Abnormal tau, mitochondrial dysfunction, impaired axonal transport of mitochondria, and synaptic deprivation in Alzheimer's disease. *Brain Research, 1415*, 136–148.

Perry, G., et al. (2000). Mitochondrial abnormalities in Alzheimer disease. *Neurobiology of Aging, 21*, 213.

Eckert, A., Schulz, K. L., Rhein, V., & Götz, J. (2010). Convergence of amyloid-β and tau pathologies on mitochondria in vivo. *Molecular Neurobiology, 41*, 107–114.

Sinclair, P., Baranova, A., & Kabbani, N. (2021). Mitochondrial disruption by amyloid beta 42 identified by proteomics and pathway mapping. *Cell, 10*, 2380.

Guimas Almeida, C., Sadat Mirfakhar, F., Perdigão, C., & Burrinha, T. (2018). Impact of late-onset Alzheimer's genetic risk factors on beta-amyloid endocytic production. *Cellular and Molecular Life Sciences, 75*, 2577–2589.

Ebneth, A., et al. (1998). Overexpression of tau protein inhibits kinesin-dependent trafficking of vesicles, mitochondria, and endoplasmic reticulum: Implications for Alzheimer's disease. *The Journal of Cell Biology, 143*, 777–794.

Nunomura, A., et al. (2001). Oxidative damage is the earliest event in Alzheimer disease. *Journal of Neuropathology and Experimental Neurology, 60*, 759–767.

Rajmohan, R., & Reddy, P. H. (2017). Amyloid-beta and phosphorylated tau accumulations cause abnormalities at synapses of Alzheimer's disease neurons. *Journal of Alzheimer's Disease, 57*, 975–999.

Gowda, P., Hemachandra Reddy, P., & Kumar, S. (2021). Deregulated mitochondrial microRNAs in Alzheimer's disease: Focus on synapse and mitochondria. *Ageing Research Reviews, 73*, 101529. https://doi.org/10.1016/j.arr.2021.101529

Manczak, M., Calkins, M. J., & Reddy, P. H. (2011). Impaired mitochondrial dynamics and abnormal interaction of amyloid beta with mitochondrial protein Drp1 in neurons from patients with Alzheimer's disease: Implications for neuronal damage. *Human Molecular Genetics, 20*, 2495–2509.

Kandimalla, R., et al. (2016). Reduced dynamin-related protein 1 protects against phosphorylated tau-induced mitochondrial dysfunction and synaptic damage in Alzheimer's disease. *Human Molecular Genetics, 25*, 4881–4897.

Forner, S., Baglietto-Vargas, D., Martini, A. C., Trujillo-Estrada, L., & LaFerla, F. M. (2017). Synaptic impairment in Alzheimer's disease: A dysregulated symphony. *Trends in Neurosciences, 40*, 347–357.

MacDonald, M. E., et al. (1993). A novel gene containing a trinucleotide repeat that is expanded and unstable on Huntington's disease chromosomes. *Cell, 72*, 971–983.

Bates, G. P., et al. (2015). Huntington disease. *Nature Reviews Disease Primers, 1*, 1–21.

Gu, M., et al. (1996). Mitochondrial defect in Huntington's disease caudate nucleus. *Annals of Neurology, 39*, 385–389.

Reddy, P. H., Mao, P., & Manczak, M. (2009). Mitochondrial structural and functional dynamics in Huntington's disease. *Brain Research Reviews, 61*, 33–48.

Cui, L., et al. (2006). Transcriptional repression of PGC-1α by mutant huntingtin leads to mitochondrial dysfunction and neurodegeneration. *Cell, 127*, 59–69.

Puigserver, P., & Spiegelman, B. M. (2003). Peroxisome proliferator-activated receptor-γ coactivator 1α (PGC-1α): Transcriptional coactivator and metabolic regulator. *Endocrine Reviews, 24*, 78–90.

Finck, B. N., & Kelly, D. P. (2006). PGC-1 coactivators: Inducible regulators of energy metabolism in health and disease. *The Journal of Clinical Investigation, 116*, 615–622.

Carmo, C., Naia, L., Lopes, C., & Rego, A. C. (2018). Mitochondrial dysfunction in Huntington's disease. *Advances in Experimental Medicine and Biology, 1049*, 59–83. https://doi.org/10.1007/978-3-319-71779-1_3

Johri, A., Chandra, A., & Beal, M. F. (2013). PGC-1α, mitochondrial dysfunction, and Huntington's disease. *Free Radical Biology & Medicine, 62*, 37–46.

Hardiman, O., et al. (2017). Amyotrophic lateral sclerosis. *Nature Reviews Disease Primers, 3*, 17071.

Al-Chalabi, A., Van Den Berg, L. H., & Veldink, J. (2017). Gene discovery in amyotrophic lateral sclerosis: Implications for clinical management. *Nature Reviews. Neurology, 13*, 96–104.

Neumann, M., et al. (2006). Ubiquitinated TDP-43 in frontotemporal lobar degeneration and amyotrophic lateral sclerosis. *Science (80-.), 314*, 130–133.

Bede, P., et al. (2013). Cognitive changes predict functional decline in ALS: A population-based longitudinal study. *Neurology, 80*, 1590–1597.

Phukan, J., et al. (2012). The syndrome of cognitive impairment in amyotrophic lateral sclerosis: A population-based study. *Journal of Neurology, Neurosurgery, and Psychiatry, 83*, 102–108.

Lansbury, P. T., & Lashuel, H. A. (2006). A century-old debate on protein aggregation and neurodegeneration enters the clinic. *Nature, 443*, 774–779.

Muyderman, H., & Chen, T. (2014). Mitochondrial dysfunction in amyotrophic lateral sclerosis – A valid pharmacological target? *British Journal of Pharmacology, 171*, 2191–2205.

Kodavati, M., Wang, H., & Hegde, M. L. (2020). Altered Mitochondrial dynamics in motor neuron disease: An emerging perspective. *Cell, 9*, 1065.

Cozzolino, M., & Carrì, M. T. (2012). Mitochondrial dysfunction in ALS. *Progress in Neurobiology, 97*, 54–66.

Chapter 9
PTMs: A Missing Piece for Schizophrenia Studies

Caroline Brandão-Teles, Bradley J. Smith, and Victor Corasolla Carregari

Abstract One of the closest regulatory mechanisms to a cellular phenotype is post-translational modifications (PTMs), a diverse class of changes that proteins can undergo to change various physical and functional properties. PTMs hold great potential to better understand multifactorial diseases and disorders like schizophrenia. The field of PTMomics is still expanding and developing, though several modifications have already been implicated in the etiology and treatment of schizophrenia. Nonetheless, much has yet to be uncovered due to the vast number of modifications that occur on proteins. Here, some of the most well-supported arguments for PTM dysregulation in schizophrenia are raised, leaving the door open for multiple other modifications and their potential.

Keywords Post-translational modifications · Schizophrenia · Proteomics

9.1 Introduction

Schizophrenia (SCZ) is a severe psychiatric disorder that usually appears between late adolescence and early adulthood [1]. Increasing evidence suggests that neurodevelopmental dysfunction is a key to the pathogenesis of the disorder [2–4]. The neurodevelopmental hypothesis proposes that the emergence of schizophrenia symptoms arises from a cascade of neuropathological events triggered by the interaction between genetic and environmental factors, during developmental processes. As the brain takes on more complex functions, neurodevelopmental changes become apparent [1, 2, 5]. However, studies into the biological bases of SCZ are hindered by

Authors Caroline Brandão-Teles and Bradley J. Smith have equally contributed to this chapter.

C. Brandão-Teles · B. J. Smith · V. C. Carregari (✉)
Laboratory of Neuroproteomics, Department of Biochemistry and Tissue Biology, Institute of Biology, University of Campinas (UNICAMP), Campinas, SP, Brazil

V. Corasolla Carregari (ed.), *Understanding PTMs in Neurodegenerative Diseases*, Advances in Experimental Medicine and Biology 1382,
https://doi.org/10.1007/978-3-031-05460-0_9

ethical and experimental limitations of studying the brain in living patients. Previous research in SCZ has mainly been focused on identifying molecular dysfunctions in blood, cerebrospinal fluid, cell cultures, and postmortem brain tissue. Genome-wide association studies (GWASs) and transcriptomic studies have been elucidating gene candidates in schizophrenia-associated risk loci, mainly involved in neurodevelopmental processes, synaptic plasticity, and epigenetic regulation [6, 7]. However, a limitation of genetic approaches is its translatability to functional protein expression, since the abundance of transcripts only in part corresponds to protein abundance [8–10]. Therefore, proteomic approaches are being extensively taken in SCZ studies, providing considerable evidence into the pathophysiology of the disorder. One additional step of phenotypic regulation that goes beyond protein expression and enters the domain of protein post-translational modifications (PTMs) [11]. In this chapter, the most common protein post-translational modifications in SCZ and how they have advanced the literature will be reviewed.

9.2 PTMs in the Pathophysiology of Schizophrenia

According to the outdated hypothesis of a unique representative protein for every gene, the human proteome would comprise approximately 20,000 proteins [12]. However, when considering alternative splicing of RNA transcripts and PTMs, the number of proteins increases substantially. PTMs represent an important protein regulation mechanism, influencing their function, binding, localization, stability, and activity. Several biological processes, including mitochondrial fusion/fission, cell fate, and glucose metabolism, rely on PTMs as essential regulators, not only in normal conditions but also as a response to the pathological states, such as neurodegenerative diseases [13, 14].

The majority of previous studies in SCZ have investigated specific abnormalities in several neurotransmitter systems and neuronal subpopulations, not considering the neuropsychiatric disorder as a broader malfunction of cell biology. Multiple new pieces of evidence have shown that SCZ can lead to abnormalities in multiple cell types and cause mitochondrial and metabolic dysfunction [15]. At the time of this study, several groups have published about the importance of PTMs in the development of the SCZ and highlighting the hypothesized correlations with the origin of the disorder. In this chapter, we have included what we consider to be the most relevant PTMs for studying SCZ pathophysiology.

9.2.1 Phosphorylation

Phosphorylation is the most widespread PTM [16] in which a phosphoryl group is attached to a serine, threonine, or tyrosine amino acids residue. This results in alterations of the conformation, stability, localization, and activity of its target protein [17]. Phosphorylation and dephosphorylation are responsible for the activation and inhibition of a range of cellular processes, including cell cycle progression, cell growth and death, cell differentiation, cell signaling transduction, gene expression, and homeostasis [18, 19]. In fact, about 30% of all eukaryotic cellular proteins are phosphorylated at one or more residues [20]. Although the modification itself is simple, a large family of proteins are responsible for regulating its addition and removal, so much so that ~2% of the eukaryotic genome is estimated to encode kinases and phosphatases [21, 22]. Many studies have focused on protein phosphorylation, and mass spectrometry (MS) has been regarded as a strong tool for the identification of phosphorylation sites, including in SCZ research [19, 23, 24].

A study of the phosphoproteome of corpus callosum from *post mortem* brain tissue of SCZ patients revealed differentially phosphorylated proteins, mostly related to ephrin B and ciliary neurotrophic factor signaling [23]. These proteins play important roles in glial cell processes, such as signaling and cellular communication, pathways that are known to be dysregulated in schizophrenia [25]. An analysis of blood serum from SCZ patients also showed alterations in phosphorylation patterns, principally associated with acute phase response signaling and coagulation systems, despite no difference in protein levels [26]. In addition to the phosphoproteomic changes inherent to schizophrenia, antipsychotic treatment has also been seen to induce alterations in the phosphorylation state of serum proteins; without changing their overall abundance [27]. As such, alterations in protein phosphorylation hold potential not only for the etiology and progression of SCZ, but also its treatment.

Though SCZ has a strong neurodevelopmental foundation, various neurotransmitter systems are also dysregulated, including the glutamatergic system [28, 29]. Disruptions in phosphorylation levels of NMDA receptors have been reported in SCZ; decreased levels of phosphorylation of a subunit of NMDA receptors were found in the frontal cortex and hippocampus of SCZ patients [30]. This alteration seems to cause impairment of NMDAR synaptic incorporation, along with disruptions in synaptic plasticity and impaired long-term potentiation and AMPAR-mediated synaptic transmission [31]. In particular, mGluR5 was seen to be a leading cause of glutamatergic dysfunction, a protein that heavily relies on phosphorylation to modulate its activity [32].

Brain-derived neurotrophic factor (BDNF) is a protein that is phosphorylated in the brain in response to an enriched environment [33] and can in turn phosphorylate NMDAR, increasing its activity [34]. In the brains of SCZ patients, BDNF has been found at reduced levels [35]. BDNF is also regulated by other proteins in the CREB/BDNF/TrkB pathway [36], which have also been found to be differentially phosphorylated in an animal model of schizophrenia using MK-801, an NMDAR antagonist; moreover, when ω-3 polyunsaturated fatty acids were administered to these MK-801-exposed animals, both the phosphorylation levels and the cognitive phenotype were rescued [37]. Given the role of this pathway in learning and cognition, as well as synaptic and behavioral plasticity [31], disrupted phosphorylation in this pathway may potentially be associated with the cognitive impairments that are observed in SCZ.

The AKT-mTOR pathway is also crucial in neurodevelopment and synaptic plasticity, and in SCZ, this signal transduction pathway and AMPK can both generate dysregulated phosphorylation profiles [38]. In that same study, haloperidol, a commonly used antipsychotic in schizophrenia, also modified the phosphorylation profiles of one of the same proteins, S6RP, suggesting a possible therapeutic effect of antipsychotics. Phosphorylation of mTOR complexes 1 and 2 mTORC1/2 have also both been seen to be disturbed in schizophrenia [39]; these complexes regulate the actin cytoskeleton and protein translation, and as such, affecting their activity may lead to phenotypic changes seen in SCZ.

Insulin signaling, specifically in neurons, is also affected in schizophrenia, and one mechanism has been proposed to be lower phosphorylation ratios of certain signaling proteins including mTOR and p70S6K, which were even correlated to PANSS symptom severity scores, in the case of AKT phosphorylation [40]. Other peripheral phosphoproteins have been implicated in antipsychotic treatment response, though the mechanisms behind the modifications and the downstream effects of their regulation are still not entirely understood [27].

9.2.2 Glycosylation

Another important PTM involved in cell signaling, growth, and differentiation; protein folding; molecular trafficking; and modulating immune responses is glycosylation [41, 42]. In the central nervous system and the brain, glycoproteins are vital for cell-cell interactions at synapses, also regulating neural development, synaptic transmission and plasticity, and memory formation [43, 44]. Over the past 25 years, several studies have found abnormal glycosylation in SCZ during both synthesis and expression [43–45]. Abnormal levels of glycans in cerebrospinal fluid in SCZ are apparent in 95% of patients, indicating that this N-glycosylation signature could be a powerful diagnostic tool for SCZ [46]. Several studies using post-mortem brains from SCZ patients have shown abnormal glycosylation of neurotransmission, mainly glutamate and γ-aminobutyric acid (GABA) signaling [15, 45]. GWAS studies have also identified aberrant glycosylation in voltage-gated calcium channels, dopamine signaling, and complement glycoproteins [47].

9.2.3 Ubiquitination

The ubiquitin – proteasome system (UPS) is a critical protein degradation system of cytosolic, nuclear, and transmembrane proteins, reported as a canonical pathway in several neuropsychiatric disorders [48]. In SCZ, several studies have shown alterations in UPS in whole blood and multiple regions of *post mortem* brain tissue [49]. Despite the extent of UPS alterations, few studies investigated its effects on protein expression in schizophrenia.

Nonetheless, mass spectrometry analyses showed an increase in protein ubiquitination in the insoluble fraction of prefrontal cortex of schizophrenia [50] and reduced protein ubiquitination, free ubiquitin, and K48-linked ubiquitination were all found in the superior temporal gyrus of *post mortem* brain tissue samples from SCZ. Furthermore, increased levels of K63 polyubiquitination and abnormalities in E1 activases and E3 ligases have also been reported [51]. In another study, alterations in the levels of free mono-ubiquitin, ubiquitinated proteins, catalytic ubiquitination, and proteasome activities in *post-mortem* orbitofrontal cortex were reported in SCZ [52]. Compared to controls, the levels of ubiquitinated proteins were 25% higher in SCZ samples, and were also higher in erythrocytes from recent-onset and treatment-resistant SCZ [52].

9.2.4 Palmitoylation

Cysteine palmitoylation (S-palmitoylation) is a PTM where a 16-carbon, saturated fatty acid is reversibly and enzymatically attached to a cysteine residue of a target protein [53]. Palmitoylation increases protein hydrophobicity, aiding protein insertion into lipid bilayers of plasma membranes and inter-compartment shuttling of proteins. This PTM has been associated as an important regulator of neuronal development and synaptic plasticity, playing important roles in neuronal processes that include axonal development, G-protein signaling, dendritogenesis, spinogenesis, and synaptic plasticity [54–56]. As neurotransmitter receptors and other membrane proteins are carefully regulated in the brain and at synapses, modifying the palmitoylation profile of proteins can lead to detrimental effects in these pathways. Several neurodegenerative diseases, including schizophrenia, have implicated disruptions in palmitoylation processes [57]. For example, decreased levels of palmitoylation were found on some proteins in SCZ dorsolateral prefrontal cortex, including VGLUT1, Ras, and MBP [58]. In one genomic study of schizophrenia, individuals with a 22q11.2 deletion exhibited a higher risk for developing schizophrenia [53, 59] a region that encodes ZDHHC8, transmembrane palmitoyltransferase that is highly expressed in the brain [59]. When 22q11.2 deletions were studied both *in vivo* and *in vivo,* the resulting impaired palmitoylation culminated in axonal growth deficiencies throughout embryogenesis as well as in terminal arborization, resulting in impairment of synaptic connectivity and spatial working memory [60]. These findings indicate possible mechanisms behind cognitive symptoms of SCZ.

9.3 Future Directions and Concluding Remarks

Despite the increase in studies of schizophrenia involving PTMs in recent years, much of the field of PTMomics has yet to be explored. Methylation, both of histones as an epigenetic regulator and of proteins as a modulator of hydrophobicity and charge, has been largely left unstudied in the context of schizophrenia. Other lipidation types have also been poorly studied in schizophrenia and its treatment, though some links have been hypothesized with N-myristoylation, prenylation,

S-palmitoylation, cholesterylation, octanoylation [15]. Protein acetylation has been mostly focused on histone modification, though tubulin acetylation is a promising target for understanding cytoskeletal dysregulations in neurodegenerative diseases [61]. Metabolism-associated modifications such as succinylation and malonylation, despite being the fourth and sixth most frequent PTMs in the dbPTM databank [16], have also been largely unstudied in brain disorders, having only recently been recognized as important protein modifications in the brain [62, 63].

In addition to the benefits of understanding the regulatory effects of PTMs on proteins and how their dysregulation can lead to the phenotypic profiles of schizophrenia, PTMs also hold great potential as biomarkers and adjunctive targets to current therapies. Changes in peripheral protein PTM profiles in SCZ and antipsychotic response, such as is seen in cerebrospinal fluid protein glycosylation and blood protein ubiquitination and phosphorylation sites might be compiled to create biomarkers or biomarker panels to assist in diagnosing schizophrenia, choosing treatment options, and following up on therapy. Moreover, separating the changes in PTM profiles due to the therapeutic effects of an antipsychotic from those that lead to undesired side effects may also assist in improving treatment options and reducing the negative effects of treatment.

Though many methods for proteomic studies can be modified to identify PTMs, oftentimes protocols must be modified to include higher-resolution instruments, enrichment steps, or complete overhauls of the experimental design. Samples for PTM co-localization studies, for example, cannot be prepared in the same way as for shotgun proteomics. Making studying PTMs even more difficult, not every modification can be enriched, or can only be done with pan-antibody-based enrichment, a costly procedure. Nonetheless, identifying alterations at the PTM level may lead to critical discoveries in common mechanisms that underlie SCZ, potentially giving a reason for the disturbances in different brain regions, cell-types, and neurotransmitter systems and representing targets for novel treatment options.

References

1. Owen, M. J., & O'Donovan, M. C. (2017). Schizophrenia and the neurodevelopmental continuum: Evidence from genomics. *World Psychiatry, 16*, 227–235.
2. Murray, R. M., & Lewis, S. W. (1988). Is schizophrenia a neurodevelopmental disorder? *BMJ, 296*, 63–63.
3. Owen, M. J., O'Donovan, M. C., Thapar, A., & Craddock, N. (2011). Neurodevelopmental hypothesis of schizophrenia. *The British Journal of Psychiatry, 198*, 173–175.
4. Stachowiak, M. K., et al. (2013). Schizophrenia: A neurodevelopmental disorder—Integrative genomic hypothesis and therapeutic implications from a transgenic mouse model. *Schizophrenia Research, 143*, 367–376.
5. Rapoport, J. L., Giedd, J. N., & Gogtay, N. (2012). Neurodevelopmental model of schizophrenia: Update 2012. *Molecular Psychiatry, 17*, 1228–1238.
6. Lee, K. W., Woon, P. S., Teo, Y. Y., & Sim, K. (2012). Genome wide association studies (GWAS) and copy number variation (CNV) studies of the major psychoses: What have we learnt? *Neuroscience and Biobehavioral Reviews, 36*, 556–571.

7. Prata, D. P., Costa-Neves, B., Cosme, G., & Vassos, E. (2019). Unravelling the genetic basis of schizophrenia and bipolar disorder with GWAS: A systematic review. *Journal of Psychiatric Research, 114*, 178–207.
8. Vogel, C., & Marcotte, E. M. (2012). Insights into the regulation of protein abundance from proteomic and transcriptomic analyses. *Nature Reviews. Genetics, 13*, 227–232.
9. Liu, Y., Beyer, A., & Aebersold, R. (2016). On the dependency of cellular protein levels on mRNA abundance. *Cell, 165*, 535–550.
10. Buccitelli, C., & Selbach, M. (2020). mRNAs, proteins and the emerging principles of gene expression control. *Nature Reviews. Genetics, 21*, 630–644.
11. Nascimento, J. M., & Martins-de-Souza, D. (2015). The proteome of schizophrenia. *npj Schizophrenia, 1*, 14003.
12. Ponomarenko, E. A., et al. (2016). The size of the human proteome: The width and depth. *International Journal of Analytical Chemistry, 2016*, 7436849.
13. Klimova, N., Long, A., & Kristian, T. (2018). Significance of mitochondrial protein post-translational modifications in pathophysiology of brain injury. *Translational Stroke Research, 9*, 223–237.
14. Tarazona, O. A., & Pourquié, O. (2020). Exploring the influence of cell metabolism on cell fate through protein post-translational modifications. *Developmental Cell, 54*, 282–292.
15. Mueller, T. M., & Meador-Woodruff, J. H. (2020). Post-translational protein modifications in schizophrenia. *NPJ Schizophrenia, 6*, 5.
16. Ramazi, S., & Zahiri, J. (2021). Posttranslational modifications in proteins: Resources, tools and prediction methods. *Database, 2021*, baab012.
17. Thingholm, T. E., Jensen, O. N., & Larsen, M. R. (2009). Analytical strategies for phosphoproteomics. *Proteomics, 9*, 1451–1468.
18. Cohen, P. (2000). The regulation of protein function by multisite phosphorylation – A 25 year update. *Trends in Biochemical Sciences, 25*, 596–601.
19. Dephoure, N., Gould, K. L., Gygi, S. P., & Kellogg, D. R. (2013). Mapping and analysis of phosphorylation sites: A quick guide for cell biologists. *Molecular Biology of the Cell, 24*, 535–542.
20. Pinna, L. A., & Ruzzene, M. (1996). How do protein kinases recognize their substrates? *Biochimica et Biophysica Acta (BBA) - Molecular Cell Research, 1314*, 191–225.
21. Ubersax, J. A., & Ferrell, J. E., Jr. (2007). Mechanisms of specificity in protein phosphorylation. *Nature Reviews Molecular Cell Biology, 8*, 530–541.
22. Manning, G., Whyte, D. B., Martinez, R., Hunter, T., & Sudarsanam, S. (2002). The protein kinase complement of the human genome. *Science, 298*, 1912–1934.
23. Saia-Cereda, V. M., Cassoli, J. S., Schmitt, A., Falkai, P., & Martins-de-Souza, D. (2016). Differential proteome and phosphoproteome may impact cell signaling in the corpus callosum of schizophrenia patients. *Schizophrenia Research, 177*, 70–77.
24. Collins, M. O., Yu, L., & Choudhary, J. S. (2007). Analysis of protein phosphorylation on a proteome-scale. *Proteomics, 7*, 2751–2768.
25. Bernstein, H.-G., Steiner, J., Guest, P. C., Dobrowolny, H., & Bogerts, B. (2015). Glial cells as key players in schizophrenia pathology: Recent insights and concepts of therapy. *Schizophrenia Research, 161*, 4–18.
26. Jaros, J. A. J., et al. (2012). Protein phosphorylation patterns in serum from schizophrenia patients and healthy controls. *Journal of Proteomics, 76 Spec No*, 43–55.
27. Jaros, J. A., et al. (2015). Effects of olanzapine on serum protein phosphorylation patterns in patients with schizophrenia. *Proteomics. Clinical Applications, 9*, 907–916.
28. Uno, Y., & Coyle, J. T. (2019). Glutamate hypothesis in schizophrenia. *Psychiatry and Clinical Neurosciences, 73*, 204–215.
29. Moghaddam, B., & Javitt, D. (2012). From revolution to evolution: The glutamate hypothesis of schizophrenia and its implication for treatment. *Neuropsychopharmacology, 37*, 4–15.

30. Emamian, E. S., Karayiorgou, M., & Gogos, J. A. (2004). Decreased phosphorylation of NMDA receptor type 1 at serine 897 in brains of patients with schizophrenia. *The Journal of Neuroscience, 24*, 1561–1564.
31. Li, B., et al. (2009). NMDA receptor phosphorylation at a site affected in schizophrenia controls synaptic and behavioral plasticity. *Journal of Neuroscience, 29*, 11965–11972.
32. Wang, H.-Y., et al. (2018). mGluR5 hypofunction is integral to glutamatergic dysregulation in schizophrenia. *Molecular Psychiatry, 25*, 750–760.
33. Xu, X., Ye, L., & Ruan, Q. (2009). Environmental enrichment induces synaptic structural modification after transient focal cerebral ischemia in rats. *Experimental Biology and Medicine, 234*, 296–305.
34. Slack, S. E., Pezet, S., McMahon, S. B., Thompson, S. W. N., & Malcangio, M. (2004). Brain-derived neurotrophic factor induces NMDA receptor subunit one phosphorylation via ERK and PKC in the rat spinal cord. *The European Journal of Neuroscience, 20*, 1769–1778.
35. Takahashi, M., et al. (2000). Abnormal expression of brain-derived neurotrophic factor and its receptor in the corticolimbic system of schizophrenic patients. *Molecular Psychiatry, 5*, 293–300.
36. Wang, H., Xu, J., Lazarovici, P., Quirion, R., & Zheng, W. (2018). cAMP response element-binding protein (CREB): A possible signaling molecule link in the pathophysiology of schizophrenia. *Frontiers in Molecular Neuroscience, 11*, 255.
37. Guo, C., et al. (2020). ω-3PUFAs improve cognitive impairments through Ser133 phosphorylation of CREB upregulating BDNF/TrkB signal in schizophrenia. *Neurotherapeutics, 17*, 1271–1286.
38. Chadha, R., Alganem, K., Mccullumsmith, R. E., & Meador-Woodruff, J. H. (2021). mTOR kinase activity disrupts a phosphorylation signaling network in schizophrenia brain. *Molecular Psychiatry*. https://doi.org/10.1038/s41380-021-01135-9
39. Chadha, R., & Meador-Woodruff, J. H. (2020). Downregulated AKT-mTOR signaling pathway proteins in dorsolateral prefrontal cortex in schizophrenia. *Neuropsychopharmacology, 45*, 1059–1067.
40. Kapogiannis, D., et al. (2019). Insulin-signaling abnormalities in drug-naïve first-episode schizophrenia: Transduction protein analyses in extracellular vesicles of putative neuronal origin. *European Psychiatry, 62*, 124–129.
41. Ohtsubo, K., & Marth, J. D. (2006). Glycosylation in cellular mechanisms of health and disease. *Cell, 126*, 855–867.
42. Reily, C., Stewart, T. J., Renfrow, M. B., & Novak, J. (2019). Glycosylation in health and disease. *Nature Reviews. Nephrology, 15*, 346–366.
43. Scott, H., & Panin, V. M. (2014). The role of protein N-glycosylation in neural transmission. *Glycobiology, 24*, 407–417.
44. Iqbal, S., Ghanimi Fard, M., Everest-Dass, A., Packer, N. H., & Parker, L. M. (2019). Understanding cellular glycan surfaces in the central nervous system. *Biochemical Society Transactions, 47*, 89–100.
45. Williams, S. E., Mealer, R. G., Scolnick, E. M., Smoller, J. W., & Cummings, R. D. (2020). Aberrant glycosylation in schizophrenia: A review of 25 years of post-mortem brain studies. *Molecular Psychiatry, 25*, 3198–3207.
46. Stanta, J. L., et al. (2010). Identification of N-glycosylation changes in the CSF and serum in patients with schizophrenia. *Journal of Proteome Research, 9*, 4476–4489.
47. Mealer, R. G., et al. (2020). Glycobiology and schizophrenia: A biological hypothesis emerging from genomic research. *Molecular Psychiatry, 25*, 3129–3139.
48. Cheon, S., Dean, M., & Chahrour, M. (2019). The ubiquitin proteasome pathway in neuropsychiatric disorders. *Neurobiology of Learning and Memory, 165*, 106791.
49. Luza, S., et al. (2020). The ubiquitin proteasome system and schizophrenia. *Lancet Psychiatry, 7*, 528–537.
50. Nucifora, L. G., et al. (2019). Increased protein insolubility in brains from a subset of patients with schizophrenia. *The American Journal of Psychiatry, 176*, 730–743.

51. Rubio, M. D., Wood, K., Haroutunian, V., & Meador-Woodruff, J. H. (2013). Dysfunction of the ubiquitin proteasome and ubiquitin-like systems in schizophrenia. *Neuropsychopharmacology, 38*, 1910–1920.
52. Bousman, C. A., et al. (2019). Elevated ubiquitinated proteins in brain and blood of individuals with schizophrenia. *Scientific Reports, 9*, 2307.
53. Hornemann, T. (2015). Palmitoylation and depalmitoylation defects. *Journal of Inherited Metabolic Disease, 38*, 179–186.
54. el-Husseini, A. E.-D., & Bredt, D. S. (2002). Protein palmitoylation: A regulator of neuronal development and function. *Nature Reviews. Neuroscience, 3*, 791–802.
55. Fukata, Y., & Fukata, M. (2010). Protein palmitoylation in neuronal development and synaptic plasticity. *Nature Reviews. Neuroscience, 11*, 161–175.
56. Globa, A. K., & Bamji, S. X. (2017). Protein palmitoylation in the development and plasticity of neuronal connections. *Current Opinion in Neurobiology, 45*, 210–220.
57. Cho, E., & Park, M. (2016). Palmitoylation in Alzheimer's disease and other neurodegenerative diseases. *Pharmacological Research, 111*, 133–151.
58. Pinner, A. L., Tucholski, J., Haroutunian, V., McCullumsmith, R. E., & Meador-Woodruff, J. H. (2016). Decreased protein S-palmitoylation in dorsolateral prefrontal cortex in schizophrenia. *Schizophrenia Research, 177*, 78–87.
59. Qin, X., Chen, J., & Zhou, T. (2020). 22q11.2 deletion syndrome and schizophrenia. *Acta Biochimica et Biophysica Sinica, 52*, 1181–1190.
60. Mukai, J., et al. (2015). Molecular substrates of altered axonal growth and brain connectivity in a mouse model of schizophrenia. *Neuron, 86*, 680–695.
61. Graziella, C., & Daniele, C. (2020). Chapter 11 – Acetylation of tubulin: A feasible protective target from neurodevelopment to neurodegeneration. In I. Gozes & J. Levine (Eds.), *Neuroprotection in autism, schizophrenia and Alzheimer's disease* (pp. 273–294). Academic Press.
62. Fadó, R., Rodríguez-Rodríguez, R., & Casals, N. (2021). The return of malonyl-CoA to the brain: Cognition and other stories. *Progress in Lipid Research, 81*, 101071.
63. Yang, Y., & Gibson, G. E. (2019). Succinylation links metabolism to protein functions. *Neurochemical Research, 44*, 2346–2359.

Chapter 10
Post-translational Modifications in Brain Diseases: A Future for Biomarkers

Licia C. Silva-Costa and Bradley J. Smith

Abstract Several omic fields have been used in the development of biomarker panels, most traditionally involving genetics and proteomics. The post-translational modification of proteins, however, is an important regulatory system of many biological processes, affecting a wide range of biochemical properties of proteins, including their binding, localization, activity, and stability. These modifications are not analyzed if not specifically searched for in proteomic workflows, making them an underrepresented source of important information in the field of biomarker research. Biomarkers can particularly benefit the diagnosis and prognosis of neurological and psychiatric diseases due to the difficulty of accessing tissue and distinguishing between multiple possible conditions. In this article, post-translational modifications in the context of brain disease are compiled, highlighting the potential that this data source holds for improving the field of medicine.

Keywords Post-translational modification · Mass spectrometry · Proteomics · Multifactorial disease

10.1 Introduction

Proteins have long been known to be modified after their translation, leading to a level of functional complexity not reachable by genetics alone. Many types of protein modifications occur *in vivo*. For example, structural modifications include not only changes in their primary structure through the splicing and cleavage of transcripts and polypeptides, but also include changes in tertiary structure by the well-known disulfide bridge. Covalent modifications of amino acid residues are highly prevalent, providing a wide range of chemical group modifications, affecting their physicochemical properties and the functional capabilities of the target protein.

L. C. Silva-Costa · B. J. Smith (✉)
Laboratory of Neuroproteomics, Department of Cell and Tissue Biology, Institute of Biology, State University of Campinas, Campinas, São Paulo, Brazil

V. Corasolla Carregari (ed.), *Understanding PTMs in Neurodegenerative Diseases*, Advances in Experimental Medicine and Biology 1382,
https://doi.org/10.1007/978-3-031-05460-0_10

Even as early as the 1970s, around 140 different covalent modifications were already known to occur [1], and the number has since surpassed 670 [2]. After decades of research, many PTMs (post-translational modifications) are well documented in regards to their effects on protein structure, function, activity, interactions, localization, export, and degradation.

The wide-ranging implications PTMs have on proteins have the potential to affect biological pathways and processes at various regulatory levels, including metabolism [3–6], the cytoskeleton [7–9], gene expression/chromatin regulation [10], and various homeostatic processes [11–15]. Since the addition and removal of PTMs can be a comparatively rapid process when considering other regulatory mechanisms such as mRNA transcription and protein translation or degradation, these pathways are therefore extremely important from a biological standpoint.

Multiple databases have been established to contain, organize, and curate the rapidly growing amount of PTM data that is continually being generated by proteomic studies, principally mass spectrometry analyses. These databases [16] have compiled a vast number of potential and confirmed sites of modification as well as the probable and confirmed effects of these PTMs; however, these data are not being used to their full potential, often due to limitations in algorithmic and processing capabilities, along with worries about the quality or trustworthiness of proteomic data from other sources [17, 18]. Nonetheless, PTMs are an important and rewarding topic to study, since their understanding can lead to a better understanding of the etiology of many diseases.

Multifactorial diseases, which often arise from a combination of many individual intrinsic and external factors, are particularly interesting in the field of PTMomics. In particular, degenerative brain disorders such as Alzheimer's and Parkinson's diseases, dementia, brain cancer, epilepsy, stroke, and psychiatric disorders are among the many complex and multifactorial diseases that can benefit from the study of PTMs. This is not only due to the wide range of affected biological targets, but also due to the difficulty of diagnosing and treating conditions that occur in such a delicate and sequestered organ.

10.2 PTM Types

Uniprot has over 670 different PTMs in its database [19] and the number grows every year. Each modification has its own mechanisms for addition and removal or reversal, when possible, and can have several enzymes involved in these processes and their regulation. Post-translational modifications can be classified into roughly four major categories: modification by the addition of a chemical group, modification by a peptide or protein, modification by amino acid residue conversion, and structural modification.

Chemical group PTMs encompass some of the most commonly recognized PTMs and include acetylation, phosphorylation, and methylation. Though functions vary widely, chemical group PTMs are often used to modify activity, binding and protein or membrane interactions, and cellular localization. Chemical PTMs can additionally function as signaling groups, as is the case with histones to recruit chromatin-modifying enzymes. Other examples of this class include the addition of acyl groups (e.g. palmitate), glycosyl groups (e.g. N-acetylglucosamine), cofactors (e.g. flavin), and other amino acids (polyglutamylation). Peptide and protein additions comprise the next major group of PTMs, including ubiquitination, neddylation, and SUMOylation. This class of modification conjugates a polypeptide to the protein of interest, often to provide a contact point for other proteins to recognize and act upon the target, affect protein stability and structure, or regulate transcription or protein binding [20–22].

Some PTMs do not consist of an addition of a chemical group, but instead can be more precisely defined as a direct modification of the amino acid residue. These include the iodination of tyrosine to form iodotyrosine, which can occur both enzymatically and non-enzymatically, the deamination of arginine to form citrulline, and the generation of hypusine from a lysine residue. This rare class of modification is still comparatively poorly understood with some functions that are hypothesized to include modulating autoimmunity, modifying tertiary structure, and affecting protein-protein interactions [23–25]. Lastly, some proteins can be modified structurally, the most well-known example being disulfide bridges between cysteine residues, which stabilizes a protein's tertiary structure, constrains protein folding outcomes [26], or stabilizes quaternary structure, commonly in antibody proteins [27, 28]. Another structural PTMs is the cleavage of peptide sequences from a protein, such as the removal of signal peptides after protein localization [29].

10.3 Regulation

Each post-translational modification can be regulated by a range of processes, and the pathway for PTM addition does not necessarily need to be the same class as the method of removal, nor must a means for addition or removal necessarily exist. The most overarching categories for PTM regulation are enzyme-based, spontaneous, and self-induced. Each method allows for a different form of PTM regulation in response to internal and external stimuli and multiple methods can exist for a single modification. The majority of modifications are those which are most often regulated by enzymes, such as acetylation, phosphorylation, and ubiquitination, as well as protein cleavage. Using enzymes to regulate PTMs gives finely tuned control over when and at which sites these modifications should occur, allowing for rapid responses to modulate protein activity, binding, localization, and other properties. Not all protein modifications occur with the help of such regulatory enzymes, however. For example, lysine succinylation and carboxylation can occur without the presence of any regulatory enzymes [30, 31], instead being correlated with succinyl-CoA [32] or carbon dioxide [33] concentration, respectively. Methionine residues can also be spontaneously oxidized into methionine sulfoxide by reactive oxygen species [34], affecting biological functions like protein binding [35]. In these cases, enzymes are still responsible for the removal or reversal of these modifications, and other cases of spontaneous, sometimes irreversible, modifications have been documented, particularly in the cardiovascular system [36–38]. Finally, a less frequent case regarding regulation involves what is called protein splicing, chemically occurring similarly to RNA splicing, in which inteins are spontaneously excised via nucleophilic attack, leaving only the exteins of a mature protein [39, 40].

10.4 PTMs in Brain Diseases

10.4.1 Neurodegenerative Disorders

The vital regulatory roles of PTMs in various biological pathways and processes thereby imply that a dysregulation of any step during the addition, removal, or recognition of PTMs could lead to any of a wide array of detrimental effects or disease states. In many brain disorders, the complete pathways behind the observed complex biological dysfunctions are not fully understood. In a study of PTM prevalence in diseases, proteins with multiple sites for PTMs were even particularly singled out due to their likelihood to be involved with serious conditions like neurodegenerative disorders, likely due to the interactions between multiple PTM types [41]. Further strengthening the importance of PTMs in biological homeostasis, DNA segments of modification sites are less likely to be the loci for rare gene variants [42] and disease-associated amino acid substitutions seem to affect known modification sites more than twice as often than point mutations that are considered functionally neutral [43]. The increased conservation of PTM sites and a higher chance of detrimental effects after changing modification sites has built a convincing base to investigate modifications in various diseases. PTMs have the ability to modulate protein-protein interactions and oligomer formation and dysregulations in these processes have been hypothesized to be behind several central nervous system (CNS) disorders.

In Parkinson's disease, aggregates of α-synuclein that form Lewy bodies [44], tau protein [45], and the phosphorylation, ubiquitination, nitration, along with protein truncation, have all been shown to contribute to its pathogenesis, as well as environmental factors that deregulate these PTMs [46–48]. In contrast, acetylation and sumoylation of certain proteins have been hypothesized to possess a protective effect [48]. While some of these PTMs seem to be directly involved with protein dimerization and solubility and the formation of (or protection against) α-synuclein aggregates [49], others seem to affect membrane binding, protein-protein interactions, and vesicle binding [50].

In Alzheimer's disease, PTM-modulated protein-protein interactions are involved in the dysfunction and tangling of microtubule-associated protein tau [51–53] and aggregates of amyloid-β [54, 55]. Both of these proteins have long been hypothesized to be involved in the pathology of the disease, contributing to the formation of insoluble aggregates when PTM states are deregulated [56]. More specifically, hyperphosphorylation of tau protein [57, 58] is seen in the neurofibrillary tangles of Alzheimer's disease, and pyroglutamate and isoaspartate modifications on amyloid-β are both increased [59], with improperly truncated amyloid-β also increasing the risk of pathogenic aggregation in conjunction with pyroglutamate [60, 61]. Several other modifications of amyloid-β and tau such as acetylation, glycation, glycosylation, methylation, nitration, oxidation, phosphorylation, prolyl-isomerization, SUMOylation, ubiquitination, are known to exist [54, 62]. Proteolytic cleavage can modulate protein activity or localization, among other functions, and the improper cleavage of organelle target sequences can lead to imbalances in protein levels, as is the case with amyloid beta cleavage dysregulating mitochondrial function in Alzheimer's disease [63]. Biomarkers that are linked to a single neurodegenerative disease can be especially important in a clinical setting to distinguish between similar diseases and assist in early diagnoses.

Some modifications, however, are involved in many neurodegenerative diseases, possibly as a response to other dysregulated processes. For example, the nitrosylation of cysteine residues is a modification that coincides with an accumulation of misfolded proteins [64, 65], and a stress-induced dysregulation of this PTM has been seen in several neurodegenerative diseases [66]. In other cases, it is not protein aggregation that is behind a disorder, but rather an imbalance in protein degradation and turnover. A wide range of PTMs affect protein turnover and assist in the removal and degradation of unwanted, accumulated, or aggregated proteins. For this reason, PTM dysregulations can contribute to the protein aggregates in several mental disorders, including Alzheimer's, Parkinson's, and Huntington's diseases, as well as amyotrophic lateral sclerosis, and some of these modifications have even been tested as therapeutic targets to reverse pathological phenotypes [67, 68]. The PTMs that are seen in multiple neurodegenerative conditions can be especially important in determining prognosis or the severity of a condition, since they most likely represent more the outcome of the disease than the underlying causes.

The immune system is also dependent on PTMs, and some modulations of modifications have been hypothesized to be behind certain aspects of neurodegenerative and psychiatric diseases. In particular, a conversion from arginine to citrulline has been extensively studied in relation to multiple sclerosis, Alzheimer's disease, and other disorders [23, 24, 69]. Proteomic studies of multiple sclerosis have also implicated methylation, deimination, phosphorylation [70], citrullination, acetylation [71], and neddylation [72] in its pathogenesis. Several modifications of cytokines have also been implicated in neuromodulation and pathogenesis, specifically neurotransmitter releases, synaptic plasticity and transmission, and neuronal excitability [73]. Cysteine and hydrogen sulfide metabolism are also both associated with protein sulfhydration, together forming an axis of neuroinflammation and neuroprotection [74].

Yet another important class of PTMs are histone modifications, a complex set of chemical groups that affect chromatin structure and protein transcription, therefore holding the potential to be associated with a wide range of disorders. Histone PTM dysregulation has indeed been seen to play a part in amyotrophic lateral sclerosis (ALS) [75]. Moreover, in models of Alzheimer's disease, Huntington's disease, and ALS, chromatin dynamics are known to be dysregulated, playing a part in neurodegenerative processes; attempts at attenuating the dysregulations even attenuated toxicity from pathogenic proteins [76].

RNA-binding proteins play a role in several regulatory pathways and have been implicated in neurodegenerative processes. Several modifications of TDP-43, a protein whose dysregulated levels in either direction lead to pathogenic phenotypes, have been associated with several neurodegenerative diseases, including ALS and frontotemporal dementia [77]. Lastly, several non-enzymatic PTMs have been tentatively associated with Alzheimer's and other neurodegenerative diseases [78]. These last groups of biomarkers that respond to the severity of several neurodegeneration can be crucial in developing more personalized treatment options in multifactorial diseases to select the most suitable therapy for a given patient.

10.4.2 Psychiatric Disorders

Compared to neurodegenerative diseases, the roles of PTMs in psychiatric diseases are even more poorly understood. Nonetheless, there is increasing evidence that PTMs explain some phenotypic features in psychiatric disorders. For example, methylated forms of precursor proteins in the arginine metabolic pathway have been implicated in the regulation of nitric oxide synthase, which may be linked to pathogenesis of major depressive disorder [79], as well as an increased risk of developing cardiovascular disease in patients with depression [80] and affecting treatment response in schizophrenia [81]. PTM dysregulation has also been suggested to be associated with stress response and resiliency [82, 83], major depressive disorder [84, 85], antidepressant response [86], neuroinflammation-induced depression [87], bipolar disorder [84], schizophrenia [88–90], and dysregulations in glutamatergic neurotransmission, related to cysteine nitrosylation and nitric oxide signaling [91]. Tests to identify these dysregulations in a given patient would greatly benefit diagnosis and treatment selection in more personalized medicine settings.

As in neurodegenerative diseases, histone modifications are also thought to play a role in psychiatric disorders. For example, in sex-associated depression-like behavior after prenatal stress, young male rats from one study showed depression-like behavior in conjunction with lower histone acetylation (H3K9ac) at tryptophan hydroxylase 2 (*Tph2*), though the effect was not observed in female rats [92]. Determining the mechanisms behind this type of dysregulation would prove beneficial to identifying and treating depression subtypes more precisely. Along these lines, measuring histone modification levels has been suggested to be an important tool in measuring treatment responses [93].

10.4.3 Aging and Cancer

Beyond affecting mental capacity and proper functioning of the mind, CNS, and peripheral nervous system, aging is also a risk factor for several pathologies, including neurodegenerative phenotypes. Understanding the links between these two conditions can provide insight into preventative measures and assist in distinguishing disease phenotypes from an aging biological profile, which could help direct treatment and prognosis. Some examples of biomarkers in this class are reduced levels of some of the sirtuin proteins, responsible for deacylation of a broad spectrum of proteins, and an accumulation of protein carbamylation with age, which has already been linked to cardiovascular disease [94] or protein oxidation, associated with several neurodegenerative and aging-related disorders [95]. The use of such aging-related PTM dysregulations as biomarkers could give insight into a patient's neurological state, ruling out other possible conditions and assisting medical professionals in selecting the best way to alleviate symptoms.

Brain cancer is also a particularly difficult type of cancer to diagnose and treat due to the enclosed nature of the brain and the difficulty of treating cancer on the other side of the blood-brain barrier. Therefore, understanding the causes and improving the diagnosis of brain cancer is paramount to a swift identification, as well as improvements in treatment. Protein stability, turnover, and conformation are all affected by glycosylation, a widespread modification, holding potential for therapeutic discoveries related to its dysregulation [96]. Moreover, epigenetic modifications affecting PTM-modulating proteins such as sirtuin deacylases, methyltransferases, and acyltransferases have been observed in glioblastoma, thereby affecting tumor development and progression [97]. Any PTMs dysregulations related to cancer that can be seen in the periphery would strongly help diagnose brain cancers and accompany relapses and treatment progression.

10.5 Biomarker Sources

One of the most difficult parts of biomarkers in the brain is the inability to measure them easily without invasive measures; the blood-brain barrier impedes a free flow of molecules between the central nervous system and the periphery and biopsies are largely infeasible. Nonetheless, advances have been made discovering non-invasive or minimally invasive biomarkers for neurological disorders. For example, while neurofilaments have already been proposed as biomarkers for several neurological disorders [98], tissue biopsy and CSF testing are far too invasive for widespread use.

Salivary biomarkers are especially exciting due to the ease of collection, and indeed several protein-based diagnostic biomarkers have been suggested for several neurological or neurodegenerative conditions, including autism, as well as Alzheimer's, Parkinson's and Huntington's diseases [99]. Nonetheless, very few post-translational modifications have been studied in this context. In one study, however, a proof of concept was shown when multiple proteins were found to be differentially phosphorylated in autism spectrum disorder [100].

Blood-based biomarkers are also very useful targets for psychiatric and neurological disorders, particularly in a clinical setting, since imaging-based screening is costly and time intensive and cerebrospinal fluid analysis is too invasive for application to the general population. In one comprehensive review, several blood-based biomarkers were compiled for neurodegenerative disorders [101]; yet once again, PTMs were rare, represented only by the phosphorylation and glycosylation of α-synuclein to detect Parkinson's disease [102]. The glycosylation of proteins is dysregulated in several cancers including brain cancer and can be seen in the periphery [96], making it a promising target of study for identifying such cancers, which are not feasible for biopsy.

Finally, yet another non-invasive source of biomarkers is urine, which can not only contain free protein but also structures such as extracellular vesicles. Analyses of extracellular vesicles are currently under study for the identification of certain conditions such as Parkinson's disease, suggesting the possibility of using urine for such conditions; however, no PTMs were included in that investigation [103].

10.6 Concluding Remarks

Several post-translational modifications have already been associated with individual components of disease states, treatment, and prognosis; however, in most cases, stringent measurements of large population groups have not been tested. Bringing these PTM dysregulations beyond mere coexistence holds the potential to identify biomarkers and form biomarker panels for the diagnosis, prognosis, and treatment accompaniment of a wide spread of brain diseases. To breach the laboratory-clinic barrier, more studies must be performed not only to understand the regulatory mechanisms behind changes in PTMs and prove robustness and uniqueness to a specific condition, but also to assess their presence in the periphery for minimally invasive measurement. As the field of biomarkers grows, adding PTMs to the list of targets will assist in the creation of more comprehensive and broadly applicable biomarker tests and panels for a future of personalized medicine.

References

1. Uy, R., & Wold, F. (1977). Posttranslational covalent modification of proteins. *Science, 198*, 890–896.
2. UniProt Consortium. (2021). *Controlled vocabulary of posttranslational modifications (PTM)*. UniProt Consortium.
3. Yu, J., & Auwerx, J. (2010). Protein deacetylation by SIRT1: An emerging key post-translational modification in metabolic regulation. *Pharmacological Research, 62*, 35–41.
4. Lushchak, O. V., Piroddi, M., Galli, F., & Lushchak, V. I. (2014). Aconitase post-translational modification as a key in linkage between Krebs cycle, iron homeostasis, redox signaling, and metabolism of reactive oxygen species. *Redox Report, 19*, 8–15.
5. Friso, G., & van Wijk, K. J. (2015). Posttranslational protein modifications in plant metabolism. *Plant Physiology, 169*, 1469–1487.
6. Sun, C.-F., Li, Y.-Q., & Mao, X.-M. (2020). Regulation of protein post-translational modifications on metabolism of actinomycetes. *Biomolecules, 10*, 1122.

7. Janke, C., & Kneussel, M. (2010). Tubulin post-translational modifications: Encoding functions on the neuronal microtubule cytoskeleton. *Trends in Neurosciences, 33*, 362–372.
8. Janke, C., & Chloë Bulinski, J. (2011). Post-translational regulation of the microtubule cytoskeleton: Mechanisms and functions. *Nature Reviews Molecular Cell Biology, 12*, 773–786.
9. Terman, J. R., & Kashina, A. (2013). Post-translational modification and regulation of actin. *Current Opinion in Cell Biology, 25*, 30–38.
10. Kennani, S. E., Crespo, M., Govin, J., & Pflieger, D. (2018). Proteomic analysis of histone variants and their PTMs: Strategies and pitfalls. *Proteomes, 6*, 29.
11. Seo, J., & Lee, K.-J. (2004). Post-translational modifications and their biological functions: Proteomic analysis and systematic approaches. *Journal of Biochemistry and Molecular Biology, 37*, 35–44.
12. Bode, A. M., & Dong, Z. (2004). Post-translational modification of p53 in tumorigenesis. *Nature Reviews Cancer, 4*, 793–805.
13. Choudhary, C., et al. (2009). Lysine acetylation targets protein complexes and co-regulates major cellular functions. *Science, 325*, 834–840.
14. Vucic, D., Dixit, V. M., & Wertz, I. E. (2011). Ubiquitylation in apoptosis: A post-translational modification at the edge of life and death. *Nature Reviews Molecular Cell Biology, 12*, 439–452.
15. Ahearn, I. M., Haigis, K., Bar-Sagi, D., & Philips, M. R. (2012). Regulating the regulator: Post-translational modification of RAS. *Nature Reviews Molecular Cell Biology, 13*, 39–51.
16. Cruz, E. R., Nguyen, H., Nguyen, T., & Wallace, I. S. (2019). Functional analysis tools for post-translational modification: A post-translational modification database for analysis of proteins and metabolic pathways. *The Plant Journal, 99*, 1003–1013.
17. Li, Y., & Chen, L. (2014). Big biological data: Challenges and opportunities. *Genomics, Proteomics & Bioinformatics, 12*, 187–189.
18. Leonelli, S. (2019). The challenges of big data biology. *eLife, 8*, e47381.
19. The UniProt Consortium. (2019). UniProt: A worldwide hub of protein knowledge. *Nucleic Acids Research, 47*, D506–D515.
20. Hagai, T., & Levy, Y. (2010). Ubiquitin not only serves as a tag but also assists degradation by inducing protein unfolding. *PNAS, 107*, 2001–2006.
21. Zou, T., & Zhang, J. (2021). Diverse and pivotal roles of neddylation in metabolism and immunity. *The FEBS Journal, 288*, 3884–3912.
22. Yang, Y., et al. (2017). Protein SUMOylation modification and its associations with disease. *Open Biology, 7*, 170167.
23. Valesini, G., et al. (2015). Citrullination and autoimmunity. *Autoimmunity Reviews, 14*, 490–497.
24. Alghamdi, M., et al. (2019). An overview of the intrinsic role of citrullination in autoimmune disorders. *Journal of Immunology Research, 2019*, e7592851.
25. Gobert, A. P., et al. (2020). Hypusination orchestrates the antimicrobial response of macrophages. *Cell Reports, 33*, 108510.
26. Muttathukattil, A. N., Singh, P. C., & Reddy, G. (2019). Role of disulfide bonds and topological frustration in the kinetic partitioning of lysozyme folding pathways. *The Journal of Physical Chemistry. B, 123*, 3232–3241.
27. Frangione, B., Prelli, F., Mihaesco, C., & Franklin, E. C. (1971). Interchain and intrachain disulfide bridges of a human immunoglobulin M: Detection of a unique fragment. *Proceedings of the National Academy of Sciences of the United States of America, 68*, 1547–1551.
28. Liu, H., & May, K. (2012). Disulfide bond structures of IgG molecules. *MAbs, 4*, 17–23.
29. Owji, H., Nezafat, N., Negahdaripour, M., Hajiebrahimi, A., & Ghasemi, Y. (2018). A comprehensive review of signal peptides: Structure, roles, and applications. *European Journal of Cell Biology, 97*, 422–441.
30. Jimenez-Morales, D., Adamian, L., Shi, D., & Liang, J. (2014). Lysine carboxylation: Unveiling a spontaneous post-translational modification. *Acta Crystallographica. Section D, Biological Crystallography, 70*, 48–57.

31. Li, F., et al. (2015). NADP+-IDH mutations promote hypersuccinylation that impairs mitochondria respiration and induces apoptosis resistance. *Molecular Cell, 60*, 661–675.
32. Weinert, B. T., et al. (2013). Lysine succinylation is a frequently occurring modification in prokaryotes and eukaryotes and extensively overlaps with acetylation. *Cell Reports, 4*, 842–851.
33. Park, I. S., & Hausinger, R. P. (1995). Requirement of carbon dioxide for in vitro assembly of the urease nickel metallocenter. *Science, 267*, 1156–1158.
34. Liang, X., et al. (2012). Characterization of methionine oxidation and methionine sulfoxide reduction using methionine-rich cysteine-free proteins. *BMC Biochemistry, 13*, 21.
35. Mo, J., et al. (2016). Understanding the impact of methionine oxidation on the biological functions of IgG1 antibodies using hydrogen/deuterium exchange mass spectrometry. *Analytical Chemistry, 88*, 9495–9502.
36. Chung, H. S., Wang, S.-B., Venkatraman, V., Murray, C. I., & Van Eyk, J. E. (2013). Cysteine oxidative post-translational modifications: Emerging regulation in the cardiovascular system. *Circulation Research, 112*, 382–392.
37. Liddy, K. A., White, M. Y., & Cordwell, S. J. (2013). Functional decorations: Post-translational modifications and heart disease delineated by targeted proteomics. *Genome Medicine, 5*, 20.
38. Wu, Z., Jankowski, V., & Jankowski, J. (2021). Irreversible post-translational modifications – Emerging cardiovascular risk factors. *Molecular Aspects of Medicine*, 101010. https://doi.org/10.1016/j.mam.2021.101010
39. Ventura, B. D., & Mootz, H. D. (2019). Switchable inteins for conditional protein splicing. *Biological Chemistry, 400*, 467–475.
40. Romero-Casañas, A., Gordo, V., Castro, J., & Ribó, M. (2020). Protein splicing: From the foundations to the development of biotechnological applications. In M. Vila-Perelló (Ed.), *Expressed protein ligation: Methods and protocols* (pp. 15–29). Springer. https://doi.org/10.1007/978-1-0716-0434-2_2
41. Huang, Q., et al. (2014). Human proteins with target sites of multiple post-translational modification types are more prone to be involved in disease. *Journal of Proteome Research, 13*, 2735–2748.
42. Reimand, J., Wagih, O., & Bader, G. D. (2015). Evolutionary constraint and disease associations of post-translational modification sites in human genomes. *PLoS Genetics, 11*, e1004919.
43. Li, S., Iakoucheva, L. M., Mooney, S. D., & Radivojac, P. (2009). Loss of post-translational modification sites in disease. In *Biocomputing 2010* (pp. 337–347). World Scientific. https://doi.org/10.1142/9789814295291_0036
44. Barrett, P. J., & Timothy Greenamyre, J. (2015). Post-translational modification of α-synuclein in Parkinson's disease. *Brain Research, 1628*, 247–253.
45. Zhang, X., et al. (2018). Tau pathology in Parkinson's disease. *Frontiers in Neurology, 9*, 809.
46. Emamzadeh, F. N., & Surguchov, A. (2018). Parkinson's disease: Biomarkers, treatment, and risk factors. *Frontiers in Neuroscience, 12*, 612.
47. Parnetti, L., et al. (2019). CSF and blood biomarkers for Parkinson's disease. *The Lancet Neurology, 18*, 573–586.
48. Pajarillo, E., Rizor, A., Lee, J., Aschner, M., & Lee, E. (1865). The role of posttranslational modifications of α-synuclein and LRRK2 in Parkinson's disease: Potential contributions of environmental factors. *Biochimica et Biophysica Acta (BBA) - Molecular Basis of Disease, 2019*, 1992–2000.
49. Gadhavi, J., Patel, M., Bhatia, D., & Gupta, S. (2022). Neurotoxic or neuroprotective: Post-translational modifications of α-synuclein at the cross-roads of functions. *Biochimie, 192*, 38–50.
50. Bell, R., & Vendruscolo, M. (2021). Modulation of the interactions between α-synuclein and lipid membranes by post-translational modifications. *Frontiers in Neurology, 12*, 661117.

51. Gong, C.-X., Liu, F., Grundke-Iqbal, I., & Iqbal, K. (2005). Post-translational modifications of tau protein in Alzheimer's disease. *Journal of Neural Transmission, 112*, 813–838.
52. Martin, L., Latypova, X., & Terro, F. (2011). Post-translational modifications of tau protein: Implications for Alzheimer's disease. *Neurochemistry International, 58*, 458–471.
53. Mair, W., et al. (2016). FLEXITau: Quantifying post-translational modifications of tau protein in vitro and in human disease. *Analytical Chemistry, 88*, 3704–3714.
54. Kummer, M. P., & Heneka, M. T. (2014). Truncated and modified amyloid-beta species. *Alzheimer's Research & Therapy, 6*, 28.
55. Schaffert, L.-N., & Carter, W. G. (2020). Do post-translational modifications influence protein aggregation in neurodegenerative diseases: A systematic review. *Brain Sciences, 10*, 232.
56. Polanco, J. C., et al. (2018). Amyloid-β and tau complexity—Towards improved biomarkers and targeted therapies. *Nature Reviews. Neurology, 14*, 22–39.
57. Perl, D. P. (2010). Neuropathology of Alzheimer's disease. *Mount Sinai Journal of Medicine, 77*, 32–42.
58. Dugger, B. N., & Dickson, D. W. (2017). Pathology of neurodegenerative diseases. *Cold Spring Harbor Perspectives in Biology, 9*, a028035.
59. Moro, M. L., et al. (2018). Pyroglutamate and isoaspartate modified amyloid-beta in ageing and Alzheimer's disease. *Acta Neuropathologica Communications, 6*, 3.
60. Gunn, A. P., Masters, C. L., & Cherny, R. A. (2010). Pyroglutamate-Aβ: Role in the natural history of Alzheimer's disease. *The International Journal of Biochemistry & Cell Biology, 42*, 1915–1918.
61. Bayer, T. A., & Wirths, O. (2014). Focusing the amyloid cascade hypothesis on N-truncated Abeta peptides as drug targets against Alzheimer's disease. *Acta Neuropathologica, 127*, 787–801.
62. Tapia-Rojas, C., et al. (2019). It's all about tau. *Progress in Neurobiology, 175*, 54–76.
63. Mossmann, D., et al. (2014). Amyloid-β peptide induces mitochondrial dysfunction by inhibition of preprotein maturation. *Cell Metabolism, 20*, 662–669.
64. Foster, M. W., McMahon, T. J., & Stamler, J. S. (2003). S-nitrosylation in health and disease. *Trends in Molecular Medicine, 9*, 160–168.
65. Nakamura, T., & Lipton, S. A. (2008). Emerging roles of S-nitrosylation in protein misfolding and neurodegenerative diseases. *Antioxidants & Redox Signaling, 10*, 87–102.
66. Foster, M. W., Hess, D. T., & Stamler, J. S. (2009). Protein S-nitrosylation in health and disease: A current perspective. *Trends in Molecular Medicine, 15*, 391–404.
67. Wang, K. K., Yang, Z., Sarkis, G., Torres, I., & Raghavan, V. (2017). Ubiquitin C-terminal hydrolase-L1 (UCH-L1) as a therapeutic and diagnostic target in neurodegeneration, neurotrauma and neuro-injuries. *Expert Opinion on Therapeutic Targets, 21*, 627–638.
68. Gupta, R., et al. (2021). Post-translational modifications: Regulators of neurodegenerative proteinopathies. *Ageing Research Reviews, 68*, 101336.
69. György, B., Tóth, E., Tarcsa, E., Falus, A., & Buzás, E. I. (2006). Citrullination: A posttranslational modification in health and disease. *The International Journal of Biochemistry & Cell Biology, 38*, 1662–1677.
70. Kim, J. K., et al. (2003). Multiple sclerosis: An important role for post-translational modifications of myelin basic protein in pathogenesis. *Molecular & Cellular Proteomics, 2*, 453–462.
71. Lillico, R., et al. (2018). Increased post-translational lysine acetylation of myelin basic protein is associated with peak neurological disability in a mouse experimental autoimmune encephalomyelitis model of multiple sclerosis. *Journal of Proteome Research, 17*, 55–62.
72. Kim, K., et al. (2021). Cell type-specific transcriptomics identifies neddylation as a novel therapeutic target in multiple sclerosis. *Brain, 144*, 450–461.
73. Vezzani, A., & Viviani, B. (2015). Neuromodulatory properties of inflammatory cytokines and their impact on neuronal excitability. *Neuropharmacology, 96*, 70–82.
74. Paul, B. D., Sbodio, J. I., & Snyder, S. H. (2018). Cysteine metabolism in neuronal redox homeostasis. *Trends in Pharmacological Sciences, 39*, 513–524.

75. Bennett, S. A., Tanaz, R., Cobos, S. N., & Torrente, M. P. (2019). Epigenetics in amyotrophic lateral sclerosis: A role for histone post-translational modifications in neurodegenerative disease. *Translational Research, 204*, 19–30.
76. Berson, A., Nativio, R., Berger, S. L., & Bonini, N. M. (2018). Epigenetic regulation in neurodegenerative diseases. *Trends in Neurosciences, 41*, 587–598.
77. Buratti, E. (2018). TDP-43 post-translational modifications in health and disease. *Expert Opinion on Therapeutic Targets, 22*, 279–293.
78. Harmel, R., & Fiedler, D. (2018). Features and regulation of non-enzymatic post-translational modifications. *Nature Chemical Biology, 14*, 244–252.
79. Ozden, A., Angelos, H., Feyza, A., Elizabeth, W., & John, P. (2020). Altered plasma levels of arginine metabolites in depression. *Journal of Psychiatric Research, 120*, 21–28.
80. Hess, S., et al. (2017). Decreased serum L-arginine and L-citrulline levels in major depression. *Psychopharmacology, 234*, 3241–3247.
81. Garip, B., Kayir, H., & Uzun, O. (2019). L-arginine metabolism before and after 10 weeks of antipsychotic treatment in first-episode psychotic patients. *Schizophrenia Research, 206*, 58–66.
82. Rein, T. (2020). Post-translational modifications and stress adaptation: The paradigm of FKBP51. *Biochemical Society Transactions, 48*, 441–449.
83. Lin, H.-Y., et al. (2020). SUMO E3 ligase PIAS1 is a potential biomarker indicating stress susceptibility. *Psychoneuroendocrinology, 120*, 104800.
84. Maes, M., Simeonova, D., Stoyanov, D., & Leunis, J.-C. (2019). Upregulation of the nitrosylome in bipolar disorder type 1 (BP1) and major depression, but not BP2: Increased IgM antibodies to nitrosylated conjugates are associated with indicants of leaky gut. *Nitric Oxide, 91*, 67–76.
85. Wang, Y., et al. (2021). Plasma nontargeted peptidomics discovers potential biomarkers for major depressive disorder. *Proteomics – Clinical Applications, 15*, 2000058.
86. Park, D. I., et al. (2018). Blood plasma/IgG N-glycome biosignatures associated with major depressive disorder symptom severity and the antidepressant response. *Scientific Reports, 8*, 179.
87. Rodriguez-Zas, S. L., et al. (2018). Disruption of microglia histone acetylation and protein pathways in mice exhibiting inflammation-associated depression-like symptoms. *Psychoneuroendocrinology, 97*, 47–58.
88. Ishida, Y. I., et al. (2017). Identification of an argpyrimidine-modified protein in human red blood cells from schizophrenic patients: A possible biomarker for diseases involving carbonyl stress. *Biochemical and Biophysical Research Communications, 493*, 573–577.
89. Bentea, E., Depasquale, E., Meller, J., Wen, Z., & McCullumsmith, R. E. (2018). F201. Kinase network dysregulation in schizophrenia: Implications for new treatment strategies. *Schizophrenia Bulletin, 44*, S299.
90. Chadha, R., & Meador-Woodruff, J. (2019). F179. Abnormalities in the AKT-mTOR signaling pathway in schizophrenia. *Biological Psychiatry, 85*, S282–S283.
91. Raju, K., et al. (2015). Regulation of brain glutamate metabolism by nitric oxide and S-nitrosylation. *Science Signaling, 8*, ra68.
92. Dang, S., et al. (2018). H3K9 acetylation of Tph2 involved in depression-like behavior in male, but not female, juvenile offspring rat induced by prenatal stress. *Neuroscience, 381*, 138–148.
93. Belzeaux, R., et al. (2018). Transcriptomic and epigenomic biomarkers of antidepressant response. *Journal of Affective Disorders, 233*, 36–44.
94. Xia, X., Chen, W., McDermott, J., & Han, J.-D. J. (2017). Molecular and phenotypic biomarkers of aging. *F1000Res, 6*, 860.
95. Santos, A. L., & Lindner, A. B. (2017). Protein posttranslational modifications: Roles in aging and age-related disease. *Oxidative Medicine and Cellular Longevity, 2017*, e5716409.
96. Veillon, L., Fakih, C., Abou-El-Hassan, H., Kobeissy, F., & Mechref, Y. (2018). Glycosylation changes in brain cancer. *ACS Chemical Neuroscience, 9*, 51–72.

97. Kunadis, E., Lakiotaki, E., Korkolopoulou, P., & Piperi, C. (2021). Targeting post-translational histone modifying enzymes in glioblastoma. *Pharmacology & Therapeutics, 220*, 107721.
98. Khalil, M., et al. (2018). Neurofilaments as biomarkers in neurological disorders. *Nature Reviews. Neurology, 14*, 577–589.
99. Farah, R., et al. (2018). Salivary biomarkers for the diagnosis and monitoring of neurological diseases. *Biomedical Journal, 41*, 63–87.
100. Castagnola, M., et al. (2008). Hypo-phosphorylation of salivary peptidome as a clue to the molecular pathogenesis of autism spectrum disorders. *Journal of Proteome Research, 7*, 5327–5332.
101. Ashton, N. J., et al. (2020). An update on blood-based biomarkers for non-Alzheimer neurodegenerative disorders. *Nature Reviews. Neurology, 16*, 265–284.
102. Vicente Miranda, H., et al. (2017). Posttranslational modifications of blood-derived alpha-synuclein as biochemical markers for Parkinson's disease. *Scientific Reports, 7*, 13713.
103. Wang, S., Kojima, K., Mobley, J. A., & West, A. B. (2019). Proteomic analysis of urinary extracellular vesicles reveal biomarkers for neurologic disease. *eBioMedicine, 45*, 351–361.

Index

V. Corasolla Carregari (ed.), *Understanding PTMs in Neurodegenerative Diseases*, Advances in Experimental Medicine and Biology 1382,
https://doi.org/10.1007/978-3-031-05460-0

The manufacturer's authorised representative in the EU is Springer Nature Customer Service Centre GmbH, Europaplatz 3, 69115 Heidelberg, Germany. If you have any concerns regarding our products, please contact ProductSafety@springernature.com

Printed and bound by CPI Group (UK) Ltd, Croydon, CR0 4YY
30/09/2026
02225465-0010